Two Manuals Combined int

Military First Aid

and

Combat and Operational Stress Control

Department of Defense

Pentagon
Publishing

*FIELD MANUAL
NO. 4-25.11
NAVY TACTICAL
REFERENCE
PUBLICATION
NO. 4-02.1
AIR FORCE MANUAL
NO. 44-163(I)

HEADQUARTERS
DEPARTMENT OF THE ARMY,
THE NAVY, AND THE AIR FORCE
Washington, DC, 23 December 2002

FIRST AID

TABLE OF CONTENTS

Page

*This publication supersedes FM 21-11, 27 October 1988

Page

PREFACE

This manual meets the first aid training needs of individual service members. Because medical personnel will not always be readily available, the nonmedical service members must rely heavily on their own skills and knowledge of life-sustaining methods to survive on the integrated battlefield. This publication outlines both self-aid and aid to other service members (buddy aid). More importantly, it emphasizes prompt and effective action in sustaining life and preventing or minimizing further suffering and disability. First aid is the emergency care given to the sick, injured, or wounded before being treated by medical personnel. The term *first aid* can be defined as "urgent and immediate lifesaving and other measures, which can be performed for casualties by nonmedical personnel when medical personnel are not immediately available." Nonmedical service members have received basic first aid training and should remain skilled in the correct procedures for giving first aid. This manual is directed to *all* service members. The procedures discussed apply to all types of casualties and the measures described are for use by both male and female service members.

This publication is in consonance with the following North Atlantic Treaty Organization (NATO) International Standardization Agreements (STANAGs) and American, British. Canadian, and Australian Quadripartite Standardization Agreements (QSTAGs).

TITLE	STANAG	QSTAG
Medical Training in First Aid, Basic Hygiene and Emergency Care	2122	535
First Aid Kits and Emergency Medical Care Kits	2126	
Medical First Aid and Hygiene Training in NBC Operations	2358	
First Aid Material for Chemical Injuries	2871	

These agreements are available on request, using Department of Defense (DD) Form 1425 from the Standardization Documents Order Desk, 700 Robins Avenue, Building 4, Section D, Philadelphia, Pennsylvania 19111-5094.

Unless this publication states otherwise, masculine nouns and pronouns do not refer exclusively to men.

Use of trade or brand names in this publication is for illustrative purposes only and does not imply endorsement by the Department of Defense (DOD).

The proponent for this publication is the US Army Medical Department Center and School. Submit comments and recommendations for the improvement of this publication directly to the **Commander, US Army Medical Department Center and School, ATTN: MCCS-FCD-L, 1400 East Grayson Street, Fort Sam Houston, Texas 78234-5052**.

CHAPTER 1

FUNDAMENTAL CRITERIA FOR FIRST AID

"The fate of the wounded rests in the hands of the ones who apply the first dressing."

Nicholas Senn (1898) (49th President of the American Medical Association)

1-1. General

When a nonmedical service member comes upon an unconscious or injured service member, he must accurately evaluate the casualty to determine the first aid measures needed to prevent further injury or death. He should seek medical assistance as soon as possible, but he should not interrupt the performance of first aid measures. To interrupt the first aid measures may cause more harm than good to the casualty. Remember that in a chemical environment, the service member should not evaluate the casualty until the casualty has been masked. After performing first aid, the service member must proceed with the evaluation and continue to monitor the casualty for development of conditions which may require the performance of necessary basic lifesaving measures, such as clearing the airway, rescue breathing, preventing shock, and controlling bleeding. He should continue to monitor the casualty until relieved by medical personnel.

Service members may have to depend upon their first aid knowledge and skills to save themselves (self-aid) or other service members (buddy aid/ combat lifesaver). They may be able to save a life, prevent permanent disability, or reduce long periods of hospitalization by knowing **WHAT** to do, **WHAT NOT** to do, and **WHEN** to seek medical assistance.

NOTE

The prevalence of various body armor systems currently fielded to US service members, and those in development for future fielding, may present a temporary obstacle to effective evaluation of an injured service member. You may have to *carefully remove* the body armor from the injured service member to complete the evaluation or administer first aid. Begin by removing the outer-most hard or soft body armor components (open, unfasten or cut the closures, fasteners, or straps), then remove any successive layers in the same manner. Be sure to follow other notes, cautions and warnings regarding procedures in contaminated situations and when a broken back or neck is suspected. Continue to evaluate.

1-2. Terminology

To enhance the understanding of the material contained in this publication, the following terms are used—

- *Combat lifesaver.* This is a US Army program governed by Army Regulation (AR) 350-41. The combat lifesaver is a member of a nonmedical unit selected by the unit commander for additional training beyond basic first aid procedures (referred to as *enhanced first aid*). A minimum of one individual per squad, crew, team, or equivalent-sized unit should be trained. The primary duty of this individual does not change. The additional duty of combat lifesaver is to provide enhanced first aid for injuries based on his training before the trauma specialist (military occupational specialty [MOS] 91W) arrives. The combat lifesaver's training is normally provided by medical personnel assigned, attached, or in direct support (DS) of the unit. The senior medical person designated by the commander manages the training program.

- *Trauma Specialist (US Army) or Hospital Corpsman (HM).* A medical specialist trained in emergency medical treatment (EMT) procedures and assigned or attached in support of a combat or combat support unit or marine forces.

- *Casualty evacuation.* Casualty evacuation (CASEVAC) is a term used by nonmedical units to refer to the movement of casualties aboard nonmedical vehicles or aircraft. See also the term *transported* below. Refer to FM 8-10-6 for additional information.

CAUTION

Casualties transported in this manner do not receive en route medical care.

- *Enhanced first aid (US Army).* Enhanced first aid is administered by the combat lifesaver. It includes measures, which require an additional level of training above self-aid and buddy aid, such as the initiation of intravenous (IV) fluids.

- *Medical evacuation.* Medical evacuation is the timely, efficient movement of the wounded, injured, or ill service members from the battlefield and other locations to medical treatment facilities (MTFs). Medical personnel provide en route medical care during the evacuation. Once the casualty has entered the medical stream (trauma specialist, hospital corpsman, evacuation crew, or MTF), the role of first aid in the care of the casualty ceases and the casualty becomes the responsibility of the health service support (HSS) chain. Once he has entered the HSS chain he is referred to as a *patient.*

- *First aid measures.* Urgent and immediate lifesaving and other measures, which can be performed for casualties (or performed by the casualty himself) by nonmedical personnel when medical personnel are not immediately available.

- *Medical treatment.* Medical treatment is the care and management of wounded, injured, or ill service members by medically trained (MOS-trained) HM, and area of concentration (AOC) personnel. It may include EMT, advanced trauma management (ATM), and resuscitative and surgical intervention.

- *Medical treatment facility.* Any facility established for the purpose of providing medical treatment. This includes battalion aid stations, Level II facilities, dispensaries, clinics, and hospitals.

- *Self-aid/buddy aid.* Each individual service member is trained to be proficient in a variety of specific first aid procedures. This training enables the service member or a buddy to apply immediate first aid measures to alleviate a life-threatening situation.

- *Transported.* A casualty is moved to an MTF in a nonmedical vehicle without en route care provided by a medically-trained service member (such as a Trauma Specialist or HM). First aid measures should be continually performed while the casualty is being transported. If the casualty is acquired by a dedicated medical vehicle with a medically-trained crew, the role of first aid ceases and the casualty becomes the responsibility of the HSS chain, and is then referred to as a *patient.* This method of transporting a casualty is also referred to as *CASEVAC.*

1-3. Understanding Vital Body Functions for First Aid

In order for the service member to learn to perform first aid procedures, he must have a basic understanding of what the vital body functions are and what the result will be if they are damaged or not functioning.

a. Breathing Process. All humans must have oxygen to live. Through the breathing process, the lungs draw oxygen from the air and put it into the blood. The heart pumps the blood through the body to be used by the cells that require a constant supply of oxygen. Some cells are more dependent on a constant supply of oxygen than others. For example, cells of the brain may die within 4 to 6 minutes without oxygen. Once these cells die, they are lost forever since they do not regenerate. This could result in permanent brain damage, paralysis, or death.

b. Respiration. Respiration occurs when a person inhales (oxygen is taken into the body) and then exhales (carbon dioxide [CO_2] is expelled from the body). Respiration involves the—

- *Airway.* The airway consists of the nose, mouth, throat, voice box, and windpipe. It is the canal through which air passes to and from the lungs.

- *Lungs.* The lungs are two elastic organs made up of thousands of tiny air spaces and covered by an airtight membrane. The *bronchial tree* is a part of the lungs.

- *Rib cage.* The rib cage is formed by the muscle-connected ribs, which join the spine in back, and the breastbone in front. The top part of the rib cage is closed by the structure of the neck, and the bottom part is separated from the abdominal cavity by a large dome-shaped muscle called the *diaphragm* (Figure 1-1). The diaphragm and rib muscles, which are under the control of the respiratory center in the brain, automatically *contract* and *relax*. *Contraction* increases and *relaxation* decreases the size of the rib cage. When the rib cage increases and then decreases, the air pressure in the lungs is first less and then more than the atmospheric pressure, thus causing the air to rush into and out of the lungs to equalize the pressure. This cycle of inhaling and exhaling is repeated about 12 to 18 times per minute.

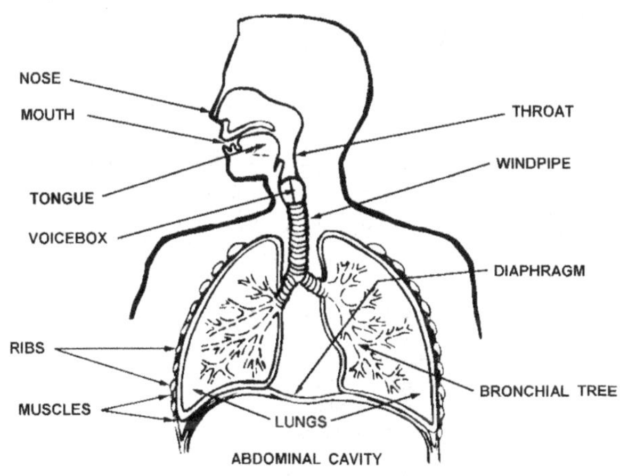

Figure 1-1. Airway, lungs, and rib cage.

c. *Blood Circulation.* The heart and the blood vessels (arteries, veins, and capillaries) circulate blood through the body tissues. The heart is divided into two separate halves, each acting as a pump. The left side pumps oxygenated blood (bright red) through the arteries into the capillaries; nutrients and oxygen pass from the blood through the walls of the capillaries into the cells. At the same time waste products and CO_2 enter the capillaries. From the capillaries the oxygen poor blood is carried through the veins to the right side of the heart and then into the lungs where it expels the CO_2 and picks up oxygen. Blood in the veins is dark red because of its low oxygen content. Blood does not flow through the veins in spurts as it does through the arteries. The entire system of the heart, blood vessels, and lymphatics is called the *circulatory system.*

(1) *Heartbeat.* The heart functions as a pump to circulate the blood continuously through the blood vessels to all parts of the body. It contracts, forcing the blood from its chambers; then it relaxes, permitting its chambers to refill with blood. The rhythmical cycle of contraction and relaxation is called the *heartbeat.* The normal heartbeat is from 60 to 80 beats per minute.

(2) *Pulse.* The heartbeat causes a rhythmical expansion and contraction of the arteries as it forces blood through them. This cycle of expansion and contraction can be felt (monitored) at various points in the body and is called the *pulse.* The common points for checking the pulse are at the—

- Side of the neck (*carotid*).
- Groin (*femoral*).
- Wrist (*radial*).
- Ankle (*posterior tibial*).

(*a*) *Carotid pulse.* To check the carotid pulse, feel for a pulse on the side of the casualty's neck closest to you. This is done by placing the tips of your first two fingers beside his Adam's apple (Figure 1-2).

Figure 1-2. Carotid pulse.

(*b*) *Femoral pulse.* To check the femoral pulse, press the tips of your first two fingers into the middle of the groin (Figure 1-3).

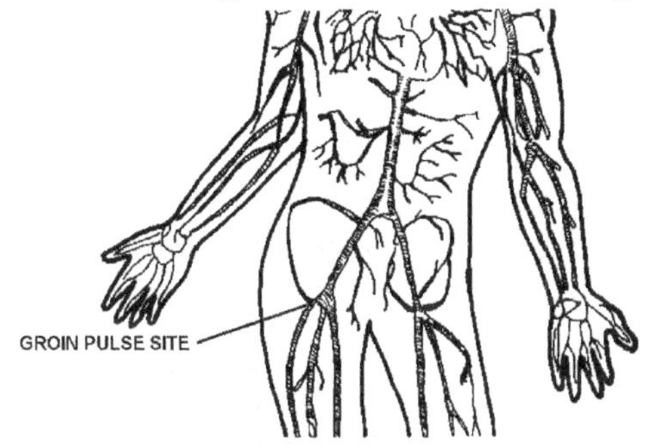

Figure 1-3. Femoral pulse.

(*c*) *Radial pulse.* To check the radial pulse, place your first two fingers on the thumb side of the casualty's wrist (Figure 1-4).

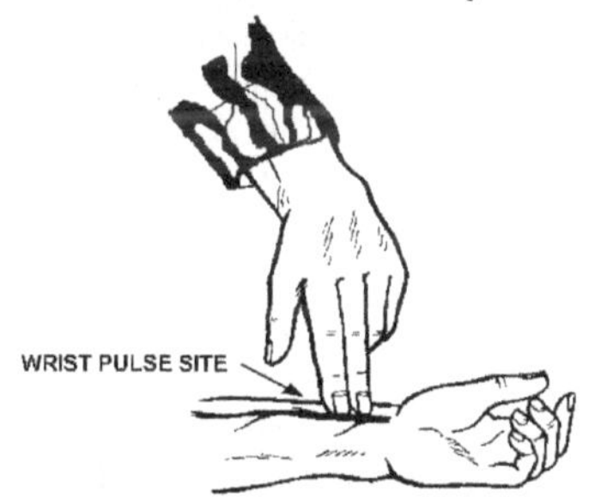

Figure 1-4. Radial pulse.

(*d*) *Posterior tibial pulse.* To check the posterior tibial pulse, place your first two fingers on the inside of the ankle (Figure 1-5).

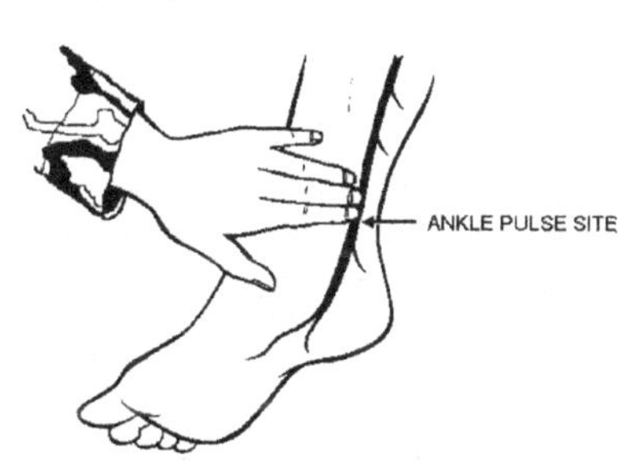

Figure 1-5. Posterior tibial pulse.

NOTE

DO NOT use your thumb to check a casualty's pulse because you may confuse the beat of your pulse with that of the casualty.

1-4. Adverse Conditions

a. Lack of Oxygen. Human life cannot exist without a continuous intake of oxygen. Lack of oxygen rapidly leads to death. First aid involves knowing how to open the airway and restore breathing.

b. Bleeding. Human life cannot continue without an adequate volume of blood circulating through the body to carry oxygen to the tissues. An important first aid measure is to stop the bleeding to prevent the loss of blood.

c. Shock. Shock means there is an inadequate blood flow to the vital tissues and organs. Shock that remains uncorrected may result in death even though the injury or condition causing the shock would not otherwise be fatal. Shock can result from many causes, such as loss of blood, loss of fluid from deep burns, pain, and reaction to the sight of a wound or blood. First aid includes preventing shock, since the casualty's chances of survival are much greater if he does not develop shock. Refer to paragraphs 2-21 through 2-24 for a further discussion of shock.

d. Infection. Recovery from a severe injury or a wound depends largely upon how well the injury or wound was initially protected. Infections result from the multiplication and growth (spread) of harmful microscopic organisms (sometimes referred to as germs). These harmful microscopic organisms are in the air, water, and soil, and on the skin and clothing. Some of these organisms will immediately invade (contaminate) a break in the skin or an open wound. The objective is to keep wounds clean and free of these organisms. A good working knowledge of basic first aid measures also includes knowing how to dress a wound to avoid infection or additional contamination.

1-5. Basics of First Aid

Most injured or ill service members are able to return to their units to fight or support primarily because they are given appropriate and timely first aid followed by the best medical care possible. Therefore, all service members must remember the basics.

- Check for **BREATHING**: Lack of oxygen intake (through a compromised airway or inadequate breathing) can lead to brain damage or death in very few minutes.

- Check for **BLEEDING**: Life cannot continue without an adequate volume of blood to carry oxygen to tissues.

- Check for **SHOCK**: Unless shock is prevented, first aid performed, and medical treatment provided, death may result even though the injury would not otherwise be fatal.

1-6. Evaluating a Casualty

a. The time may come when you must instantly apply your knowledge of first aid measures. This could occur during combat operations, in training situations, or while in a nonduty status. Any service member observing an unconscious and/or ill, injured, or wounded person must carefully and skillfully evaluate him to determine the first aid measures required to prevent further injury or death. He should seek help from medical personnel as soon as possible, but must not interrupt his evaluation of the casualty or fail to administer first aid measures. A second service member may be sent to find medical help. One of the cardinal principles for assisting a casualty is that you (the initial rescuer) must continue the evaluation and first aid measures, as the tactical situation permits, until another individual relieves you. If, during any part of the evaluation, the casualty exhibits the conditions (such as shock) for which the service member is checking, the service member must stop the evaluation and immediately administer first aid. In a chemical environment, the service member should not evaluate the casualty until both the individual and the casualty have been masked. If it is suspected that a nerve agent was used, administer the casualty's own nerve agent antidote autoinjector. After providing first aid, the service member must proceed with the evaluation and continue to monitor the casualty for further complications until relieved by medical personnel.

WARNING

Do not use your own nerve agent antidote autoinjector on the casualty.

NOTE

Remember, when evaluating and/or administering first aid to a casualty, you should seek medical aid as soon as possible. **DO NOT** stop first aid measures, but if the situation allows, send another service member to find medical aid.

b. To evaluate a casualty, perform the following steps:

(1) *Check the casualty for responsiveness.* This is done by gently shaking or tapping him while calmly asking, "Are you OK?" Watch for a response. If the casualty does not respond, go to step (2). If the casualty responds, continue with the evaluation.

(*a*) If the casualty is conscious, ask him where he feels different than usual or where it hurts. Ask him to identify the location of pain if he can, or to identify the area in which there is no feeling.

(*b*) If the casualty is conscious but is choking and cannot talk, stop the evaluation and begin first aid measures. Refer to paragraphs 2-10 and 2-11 for specific information on opening the airway.

WARNING

If a broken back or neck is suspected, do not move the casualty unless his life is in immediate danger (such as close to a burning vehicle). Movement may cause permanent paralysis or death.

(2) *Check for breathing.* (Refer to paragraph 2-6 for this procedure.)

(*a*) If the casualty is breathing, proceed to step (3).

(*b*) If the casualty is not breathing, stop the evaluation and begin first aid measures to attempt to ventilate the casualty. Attempt to open the airway, if an airway obstruction is apparent, clear the airway obstruction, then ventilate (see paragraphs 2-10 and 2-11).

(*c*) After successfully ventilating the casualty, proceed to step (3).

(3) *Check for pulse.* (Refer to paragraph 1-3*c*(2) for specific methods.) If a pulse is present and the casualty is breathing, proceed to step (4).

(*a*) If a pulse is present, but the casualty is still not breathing, start rescue breathing.

(*b*) If a pulse is not present, seek medical personnel for help.

(4) *Check for bleeding*. Look for spurts of blood or blood-soaked clothes. Also check for *both* entry and exit wounds. If the casualty is bleeding from an open wound, stop the evaluation and begin first aid procedures as follows for a—

(*a*) Wound of the arm or leg (refer to paragraphs 2-16 through 2-18 for information on putting on a field or pressure dressing).

(*b*) Partial or complete amputation, apply dressing (refer to paragraph 2-16 to 2-18) and then apply tourniquet if bleeding is not stopped (refer to paragraph 2-20 for information on putting on a tourniquet).

(*c*) Open head wound (refer to paragraph 3-10 for information on applying a dressing to an open head wound).

(*d*) Open chest wound (refer to paragraph 3-5 for information on applying a dressing to an open chest wound).

(*e*) Open abdominal wound (refer to paragraph 3-7 for information on applying a dressing to an open abdominal wound).

WARNING

In a chemically contaminated area, do not expose the wounds. Apply field dressing and then pressure dressing over wound area as needed.

(5) *Check for shock*. (Refer to paragraph 2-24 for first aid measures for shock.) If the signs and symptoms of shock are present, stop the evaluation, and begin first aid measures immediately. The following are the nine signs and symptoms of shock.

(*a*) Sweaty but cool skin (clammy skin).

(*b*) Paleness of skin. (In dark-skinned service members look for a grayish cast to the skin.)

(*c*) Restlessness or nervousness.

(*d*) Thirst.

(*e*) Loss of blood (bleeding).

(*f*) Confusion (does not seem aware of surroundings).

(*g*) Faster than normal breathing rate.

(*h*) Blotchy or bluish skin, especially around the mouth.

(*i*) Nausea or vomiting.

WARNING

Leg fractures must be splinted before elevating the legs as a first aid measure for shock.

(6) *Check for fractures*.

(*a*) Check for the following signs and symptoms of a back or neck injury and perform first aid procedures as necessary.

- Pain or tenderness of the back or neck area.
- Cuts or bruises on the back or neck area.
- Inability of a casualty to move or decreased sensation to extremities (paralysis or numbness).
  - Ask about ability to move (paralysis).
  - Touch the casualty's arms and legs and ask whether he can feel your hand (numbness).
- Unusual body or limb position.

(*b*) Immobilize any casualty suspected of having a back or neck injury by doing the following:

- Tell the casualty not to move.
- If a back injury is suspected, place padding (rolled or folded to conform to the shape of the arch) under the natural arch of the casualty's back. (For example, a blanket/poncho may be used as padding.)

WARNING

Do not move casualty to place padding.

- If a neck injury is suspected, immediately immobilize (manually) the head and neck. Place a roll of cloth under the casualty's neck, and put weighted boots (filled with dirt or sand) or rocks on both sides of his head.

(*c*) Check the casualty's arms and legs for open or closed fractures.

- Check for *open* fractures by looking for—
  - Bleeding.
  - Bones sticking through the skin.
  - Check for pulse.
- Check for *closed* fractures by looking for—
  - Swelling.
  - Discoloration.
  - Deformity.
  - Unusual body position.
  - Check for pulse.

(*d*) Stop the evaluation and begin first aid measures if a fracture to an arm or leg is suspected. Refer to Chapter 4 for information on splinting a suspected fracture.

(*e*) Check for signs/symptoms of fractures of other body areas (for example, shoulder or hip) and provide first aid as necessary.

(7) *Check for burns*. Look carefully for reddened, blistered, or charred skin; also check for singed clothing. If burns are found, stop the evaluation and begin first aid procedures. Refer to paragraph 3-9 for information on giving first aid for burns.

NOTE

Burns to the upper torso and face may cause respiratory complications. When evaluating the casualty, look for singed nose hair, soot around the nostrils, and listen for abnormal breath sounds or difficulty breathing.

(8) *Check for possible head injury.*

(*a*) Look for the following signs and symptoms:

- Unequal pupils.
- Fluid from the ear(s), nose, mouth, or injury site.
- Slurred speech.
- Confusion.
- Sleepiness.
- Loss of memory or consciousness.
- Staggering in walking.
- Headache.
- Dizziness.
- Nausea or vomiting.
- Paralysis.
- Convulsions or twitches.
- Bruising around the eyes and behind the ears.

(*b*) If a head injury is suspected, continue to watch for signs which would require performance of rescue breathing, first aid measures for shock, or control of bleeding; seek medical aid. Refer to paragraph 3-10 for information on first aid measures for head injuries.

CHAPTER 2

BASIC MEASURES FOR FIRST AID

2-1. General

Several conditions that require immediate attention are an inadequate airway, lack of breathing, and excessive loss of blood (circulation). A casualty without a clear airway or who is not breathing may die from lack of oxygen. Excessive loss of blood may lead to shock, and shock can lead to death; therefore, you must act immediately to control the loss of blood. All wounds are considered to be contaminated, since infection-producing organisms (germs) are always present on the skin and clothing, and in the soil, water, and air. Any missile or instrument (such as a bullet, shrapnel, knife, or bayonet) causing a wound pushes or carries the germs into that wound. Infection results as these organisms multiply. That a wound is contaminated does not lessen the importance of protecting it from further contamination. You must dress and bandage a wound as soon as possible to prevent further contamination.

NOTE

It is also important that you attend to any airway, breathing, or bleeding problems **IMMEDIATELY** because these problems, if left unattended, may become life threatening.

Section I. OPEN THE AIRWAY AND RESTORE BREATHING

2-2. Breathing Process

All humans must have oxygen to live. Through the breathing process, the lungs draw oxygen from the air and put it into the blood. The heart pumps the blood through the body to be used by the cells that require a constant supply of oxygen. Some cells are more dependent on a constant supply of oxygen than others. For example, cells of the brain may die within 4 to 6 minutes without oxygen. Once these cells die, they are lost forever since they do not regenerate. This could result in permanent brain damage, paralysis, or death.

2-3. Assessment of and Positioning the Casualty

a. **CHECK** for responsiveness (Figure 2-1A)—establish whether the casualty is conscious by gently shaking him and asking, “Are you OK?”

b. **CALL** for help (Figure 2-1B).

c. **POSITION** the unconscious casualty so that he is lying on his back and on a firm surface (Figure 2-1C).

> **WARNING**
>
> **If the casualty is lying on his chest (prone position), cautiously roll the casualty as a unit so that his body does not twist (which may further complicate a back, neck, or spinal injury).**

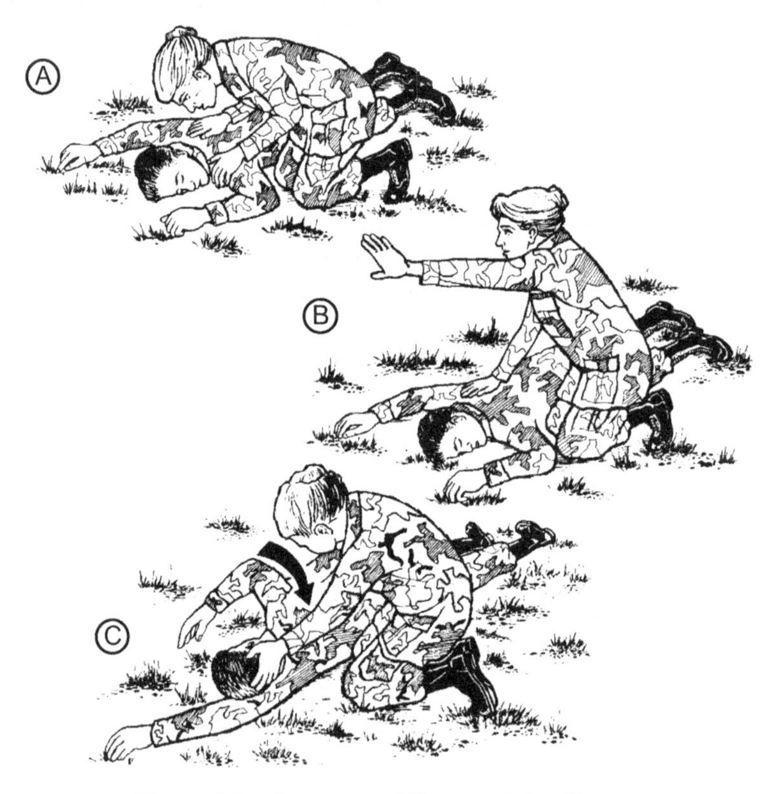

Figure 2-1. Assessment (Illustrated A—C).

(1) Straighten the casualty's legs. Take the casualty's arm that is nearest to you and move it so that it is straight and above his head. Repeat the procedure for the other arm.

(2) Kneel beside the casualty with your knees near his shoulders (leave space to roll his body) (Figure 2-1B). Place one hand behind his head and neck for support. With your other hand, grasp the casualty under his far arm (Figure 2-1C).

(3) Roll the casualty towards you using a steady, even pull. His head and neck should stay in line with his back.

(4) Return the casualty's arms to his side. Straighten his legs. Reposition yourself so that you are now kneeling at the level of the casualty's shoulders. However, if a neck injury is suspected and the jaw-thrust technique will be used, kneel at the casualty's head, looking towards his feet.

2-4. Opening the Airway of an Unconscious or Not Breathing Casualty

The tongue is the single most common cause of an airway obstruction (Figure 2-2). In most cases, simply using the head-tilt/chin-lift technique can clear the airway. This action pulls the tongue away from the air passage in the throat (Figure 2-3).

Figure 2-2. Airway blocked by tongue.

Figure 2-3. Airway opened by extending neck.

a. Call for help and then position the casualty. Move (roll) the casualty onto his back (Figure 2-1C). (Refer to paragraph 2-3*c* for information on positioning the casualty.)

NOTE

Perform finger sweep. If foreign material or vomitus is visible in the mouth, it should be removed, but do not spend an excessive amount of time doing so.

b. Open the airway using the jaw-thrust or head-tilt/chin-lift technique.

CAUTION

The head-tilt/chin-lift technique is an important procedure in opening the airway; however, use extreme care because excess force in performing this maneuver may cause further spinal injury. In a casualty with a suspected neck injury or severe head trauma, the safest approach to opening the airway is the jaw-thrust technique because in most cases it can be accomplished without extending the neck.

(1) *Perform the jaw-thrust technique.* The jaw-thrust may be accomplished by the rescuer grasping the angles of the casualty's lower jaw and lifting with both hands, one on each side, displacing the jaw forward and up (Figure 2-4). The rescuer's elbows should rest on the surface on which the casualty is lying. If the lips close, the lower lip can be retracted with the thumb. If mouth-to-mouth breathing is necessary, close the nostrils by placing your cheek tightly against them. The head should be carefully supported without tilting it backwards or turning it from side to side. If this is unsuccessful, the head should be tilted back very slightly. The jaw-thrust is the safest first approach to opening the airway of a casualty who has a suspected neck injury because in most cases it can be accomplished without extending the neck.

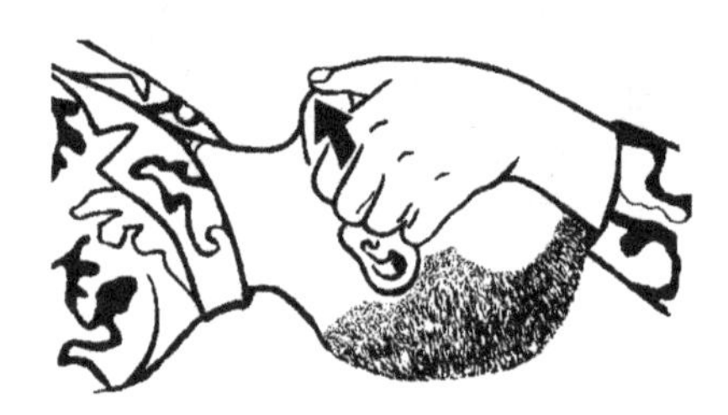

Figure 2-4. Jaw-thrust technique of opening airway.

(2) *Perform the head-tilt/chin-lift technique*. Place one hand on the casualty's forehead and apply firm, backward pressure with the palm to tilt the head back. Place the fingertips of the other hand under the bony part of the lower jaw and lift, bringing the chin forward. The thumb should not be used to lift the chin (Figure 2-5).

NOTE

The fingers should not press deeply into the soft tissue under the chin because the airway may be obstructed.

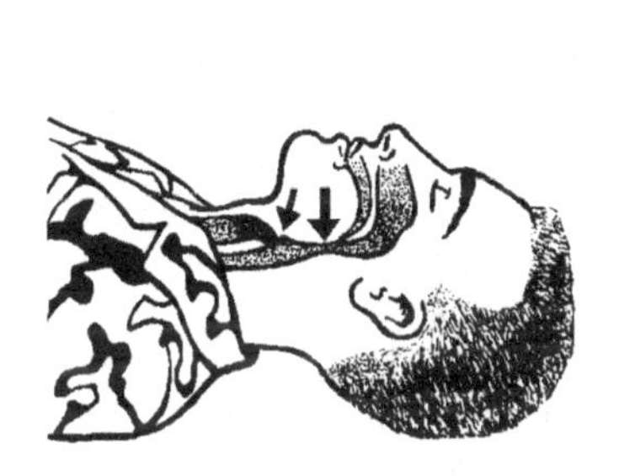

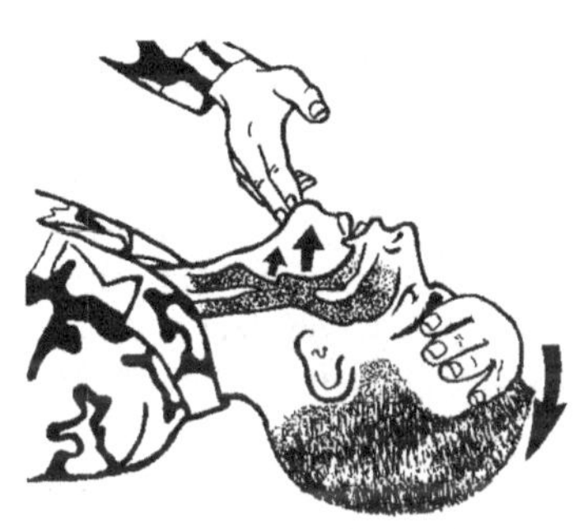

Figure 2-5. Head-tilt/chin-lift technique of opening airway.

(3) *Check for breathing (while maintaining an airway).* After establishing an open airway, it is important to maintain that airway in an open position. Often the act of just opening and maintaining the airway will allow the casualty to breathe properly. Once the rescuer uses one of the techniques to open the airway (jaw-thrust or head-tilt/chin-lift), he should maintain that head position to keep the airway open. Failure to maintain the open airway will prevent the casualty from receiving an adequate supply of oxygen. Therefore, while maintaining an open airway the rescuer should check for breathing by observing the casualty's chest and performing the following actions within 3 to 5 seconds:

(*a*) **LOOK** for the chest to rise and fall.

(*b*) **LISTEN** for air escaping during exhalation by placing your ear near the casualty's mouth.

(*c*) **FEEL** for the flow of air on your cheek (see Figure 2-6).

(*d*) **PERFORM** rescue breathing if the casualty does not resume breathing spontaneously.

NOTE

If the casualty resumes breathing, monitor and maintain the open airway. He should be transported to an MTF, as soon as practical.

2-5. Rescue Breathing (Artificial Respiration)

a. If the casualty does not promptly resume adequate spontaneous breathing after the airway is open, rescue breathing (artificial respiration) must be started. Be calm! Think and act quickly! The sooner you begin rescue breathing, the more likely you are to restore the casualty's breathing. If you are in doubt whether the casualty is breathing, give artificial respiration, since it can do no harm to a person who is breathing. If the casualty is breathing, you can feel and see his chest move. If the casualty is breathing, you can feel and hear air being expelled by putting your hand or ear close to his mouth and nose.

b. There are several methods of administering rescue breathing. The mouth-to-mouth method is preferred; however, it cannot be used in all situations. If the casualty has a severe jaw fracture or mouth wound or his jaws are tightly closed by spasms, use the mouth-to-nose method.

2-6. Preliminary Steps—All Rescue Breathing Methods

a. Establish unresponsiveness. Call for help. Turn or position the casualty.

b. Open the airway.

c. Check for breathing by placing your ear over the casualty's mouth and nose, and looking toward his chest.

(1) **LOOK** for rise and fall of the casualty's chest (Figure 2-6).

(2) **LISTEN** for sounds of breathing.

(3) **FEEL** for breath on the side of your face. If the chest does not rise and fall and no air is exhaled, then the casualty is not breathing.

(4) **PERFORM** rescue breathing if the casualty is not breathing.

NOTE

Although the rescuer may notice that the casualty is making respiratory efforts, the airway may still be obstructed and opening the airway may be all that is needed. If the casualty resumes breathing, the rescuer should continue to maintain an open airway.

Figure 2-6. Check for breathing.

2-7. Mouth-to-Mouth Method

In this method of rescue breathing, you inflate the casualty's lungs with air from your lungs. This can be accomplished by blowing air into the person's mouth. The mouth-to-mouth rescue breathing method is performed as follows:

a. If the casualty is not breathing, place your hand on his forehead, and pinch his nostrils together with the thumb and index finger of this hand. Let this same hand exert pressure on his forehead to maintain the backward head tilt and maintain an open airway. With your other hand, keep your fingertips on the bony part of the lower jaw near the chin and lift (Figure 2-7).

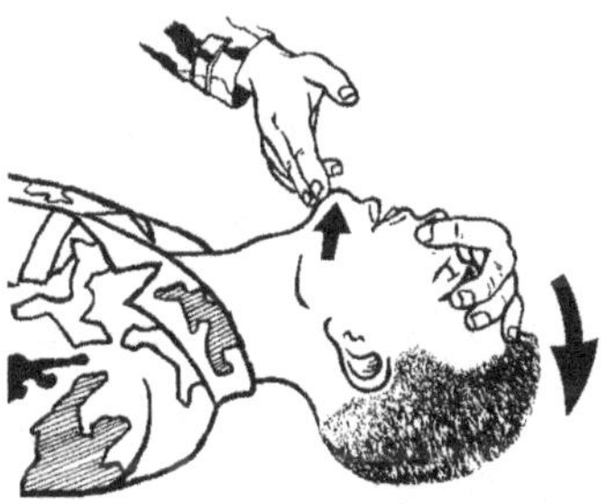

Figure 2-7. Head tilt/chin lift.

NOTE

If you suspect the casualty has a neck injury and you are using the jaw-thrust technique, close the nostrils by placing your cheek tightly against them.

b. Take a deep breath and place your mouth (in an airtight seal) around the casualty's mouth (Figure 2-8). (If the injured person is small, cover both his nose and mouth with your mouth, sealing your lips against the skin of his face.)

Figure 2-8. Rescue breathing.

c. Blow two full breaths into the casualty's mouth (1 to 1 1/2 seconds per breath), taking a breath of fresh air each time before you blow. Watch out of the corner of your eye for the casualty's chest to rise. If the chest rises, sufficient air is getting into the casualty's lungs. Therefore, proceed as described in step (1). If the chest does not rise, do the following (*a*, *b*, and *c* below) and then attempt to ventilate again.

(1) Take corrective action immediately by reestablishing the airway. Make sure that air is not leaking from around your mouth or out of the casualty's pinched nose.

(2) Reattempt to ventilate.

(3) If the chest still does not rise, take the necessary action to open an obstructed airway (paragraph 2-10).

NOTE

If the initial attempt to ventilate the casualty is unsuccessful, reposition the casualty's head and repeat rescue breathing. Improper chin and head positioning is the most common cause of difficulty with ventilation. If the casualty cannot be ventilated after repositioning the head, proceed with foreign-body airway obstruction maneuvers (see paragraph 2-10).

(4) After giving two slow breaths, which cause the chest to rise, attempt to locate a pulse on the casualty. Feel for a pulse on the side of the casualty's neck closest to you by placing the first two fingers (index and middle fingers) of your hand on the groove beside the casualty's Adam's apple (carotid pulse) (Figure 2-9). (Your thumb should not be used for pulse taking because you may confuse your pulse beat with that of the casualty.) Maintain the airway by keeping your other hand on the casualty's forehead. Allow 5 to 10 seconds to determine if there is a pulse.

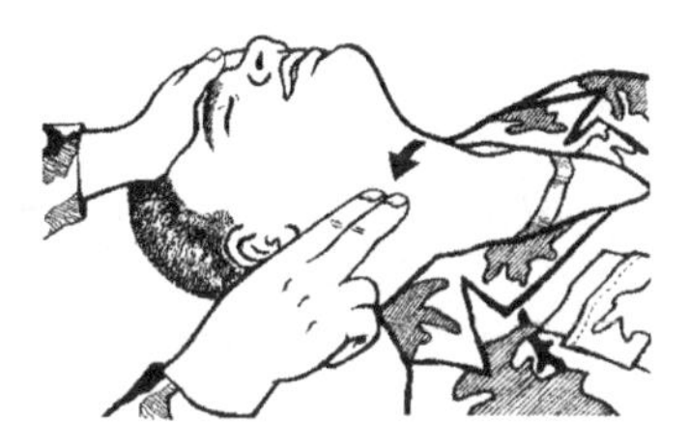

Figure 2-9. Placement of fingers to detect pulse.

(*a*) If signs of circulation are present and a pulse is found and the casualty is breathing—**STOP**; allow the casualty to breathe on his own. If possible, keep him warm and comfortable.

(*b*) If a pulse is found and the casualty is not breathing, continue rescue breathing.

(*c*) If a pulse is not found, seek medically trained personnel for help as soon as possible.

2-8. Mouth-to-Nose Method

Use this method if you cannot perform mouth-to-mouth rescue breathing because the casualty has a severe jaw fracture or mouth wound or his jaws are tightly closed by spasms. The mouth-to-nose method is performed in the same way as the mouth-to-mouth method except that you blow into the nose while you hold the lips closed with one hand at the chin. You then remove your mouth to allow the casualty to exhale passively. It may be necessary to separate the casualty's lips to allow the air to escape during exhalation.

2-9. Heartbeat

If a casualty's heart stops beating, you must immediately seek medical help. **SECONDS COUNT!** Stoppage of the heart is soon followed by cessation of respiration unless it has occurred first. Be calm! Think and act! When a casualty's heart has stopped, there is no pulse at all; the person is unconscious and limp, and the pupils of his eyes are open wide. When evaluating a casualty or when performing the preliminary steps of rescue breathing, feel for a pulse. If you DO NOT detect a pulse, seek medical help.

2-10. Airway Obstructions

In order for oxygen from the air to flow to and from the lungs, the upper airway must be unobstructed.

a. Upper airway obstructions often occur because—

(1) The casualty's tongue falls back into his throat while he is unconscious. The tongue *falls back* and *obstructs* the airway, it is not swallowed by the casualty.

NOTE

Ensure the correct positioning and maintenance of the open airway for an injured or unconscious casualty.

(2) Foreign bodies become lodged in the throat. These obstructions usually occur while eating. Choking on food (usually meat) is associated with—

- Attempting to swallow large pieces of poorly chewed food.
- Drinking alcohol.
- Slipping dentures.

(3) The contents of the stomach are regurgitated and may block the airway.

(4) Blood clots may form as a result of head and facial injuries.

b. Upper airway obstruction may cause either partial or complete airway blockage.

(1) *Partial airway obstruction.* The casualty may still have an air exchange. A good air exchange means that the casualty can cough

forcefully, though he may be wheezing between coughs. You, the rescuer, should not interfere, and should encourage the casualty to cough up the object obstructing his airway on his own. A poor air exchange may be indicated by weak coughing with a high pitched noise between coughs. Further, the casualty may show signs of shock (paragraph 1-6*b*[5]) indicating a need for oxygen. You should assist the casualty and treat him as though he had a complete obstruction.

(2) *Complete airway obstruction.* A complete obstruction (no air exchange) is indicated if the casualty cannot speak, breathe, or cough at all. He may be clutching his neck and moving erratically. In an unconscious casualty, a complete obstruction is also indicated if after opening his airway you cannot ventilate him.

2-11. Opening the Obstructed Airway—Conscious Casualty

Clearing a conscious casualty's airway obstruction can be performed with the casualty either standing or sitting and by following a relatively simple procedure.

> **WARNING**
>
> **Once an obstructed airway occurs, the brain will develop an oxygen deficiency resulting in unconsciousness. Death will follow rapidly if breathing is not promptly restored.**

a. Ask the casualty if he can speak or if he is choking. Check for the universal choking sign (Figure 2-10).

Figure 2-10. Universal sign of choking.

b. If the casualty can speak, encourage him to attempt to cough; the casualty still has a good air exchange. If he is able to speak or cough effectively, DO NOT interfere with his attempts to expel the obstruction.

c. Listen for high pitched sounds when the casualty breathes or coughs (poor air exchange). If there is poor air exchange or no breathing, CALL FOR HELP and immediately deliver manual thrusts (either an abdominal or chest thrust).

NOTE

The manual thrust with the hands centered between the waist and the rib cage is called an abdominal thrust (or Heimlich maneuver). The chest thrust (the hands are centered in the middle of the breastbone) is used only for an individual in the advanced stages of pregnancy, in the markedly obese casualty, or if there is a significant abdominal wound.

(1) Apply abdominal thrusts. This can be accomplished by using the following procedures:

(*a*) Stand behind the casualty and wrap your arms around his waist.

(*b*) Make a fist with one hand and grasp it with the other. The thumb side of your fist should be against the casualty's abdomen, in the midline and slightly above the casualty's navel, but well below the tip of the breastbone (Figure 2-11).

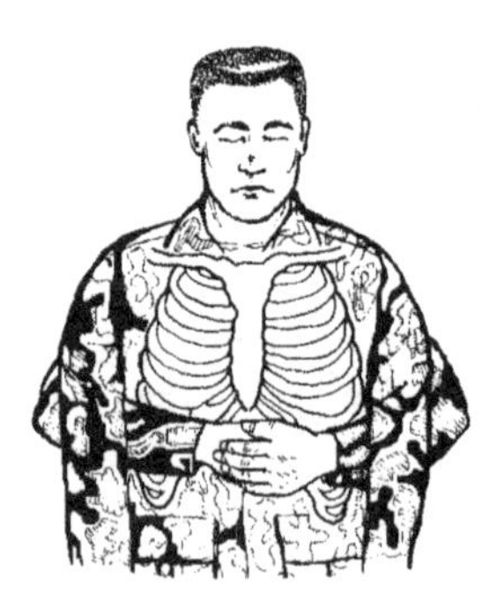

Figure 2-11. Anatomical view of abdominal thrust procedure.

(*c*) Press the fists into the abdomen with a quick backward and upward thrust (Figure 2-12).

Figure 2-12. Profile view of abdominal thrust.

(*d*) Each thrust should be a separate and distinct movement.

NOTE

Continue performing abdominal thrusts until the obstruction is expelled or the casualty becomes unresponsive.

(*e*) If the casualty becomes unresponsive, call for help as you proceed with steps to open the airway, and perform rescue breathing. (Refer to paragraph 2-7 for information on how to perform mouth-to-mouth resuscitation.)

(2) Apply chest thrusts. An alternate technique to the abdominal thrust is the chest thrust. This technique is useful when the casualty has an abdominal wound, when the casualty is pregnant, or when the casualty is so large that you cannot wrap your arms around the abdomen. To apply chest thrusts with casualty sitting or standing:

(*a*) Stand behind the casualty and wrap your arms around his chest with your arms under his armpits.

(*b*) Make a fist with one hand and place the thumb side of the fist in the middle of the breastbone (take care to avoid the tip of the breastbone and the margins of the ribs).

(*c*) Grasp the fist with the other hand and exert thrusts (Figure 2-13).

Figure 2-13. Profile view of chest thrust.

(*d*) Each thrust should be delivered slowly, distinctly, and with the intent of relieving the obstruction.

(*e*) Perform chest thrusts until the obstruction is expelled or the casualty becomes unresponsive.

(*f*) If the casualty becomes unresponsive, call for help as you proceed with steps to open the airway and perform rescue breathing.

2-12. Opening the Obstructed Airway—Casualty Lying Down or Unresponsive

The following procedures are used to expel an airway obstruction in a casualty who is lying down, who becomes unconscious, or who is found unconscious (the cause unknown):

- If a conscious casualty who is choking becomes unresponsive, call for help, open the airway, perform a finger sweep, and attempt rescue breathing (paragraphs 2-4 through 2-8). If you still cannot administer rescue breathing due to an airway blockage, then remove the airway obstruction using the procedures as in *b* below.

- If a casualty is unresponsive when you find him (the cause unknown), assess or evaluate the situation, call for help, position the casualty on his back, open the airway, establish breathlessness, and attempt to perform rescue breathing (paragraphs 2-4 through 2-8).

a. Open the airway and attempt rescue breathing (refer to paragraph 2-7 for information on how to perform mouth-to-mouth resuscitation).

b. If still unable to ventilate the casualty, perform 6 to 10 manual (abdominal or chest) thrusts.

(1) To perform the abdominal thrusts:

(*a*) Kneel astride the casualty's thighs (Figure 2-14).

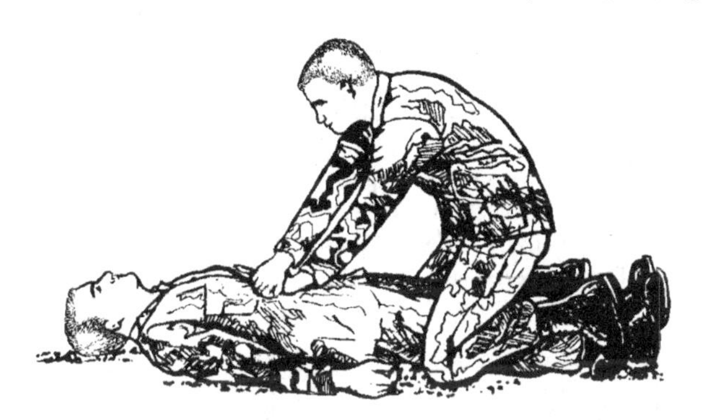

Figure 2-14. Abdominal thrust on unresponsive casualty.

(*b*) Place the heel of one hand against the casualty's abdomen (in the midline slightly above the navel but well below the tip of the breastbone). Place your other hand on top of the first one. Point your fingers toward the casualty's head.

(*c*) Press into the casualty's abdomen with a quick, forward and upward thrust. You can use your body weight to perform the maneuver. Deliver each thrust quickly and distinctly.

(*d*) Repeat the sequence of abdominal thrusts, finger sweep, and rescue breathing (attempt to ventilate) as long as necessary to remove the object from the obstructed airway.

(*e*) If the casualty's chest rises, proceed to feeling for pulse.

(2) To perform chest thrusts:

(*a*) Place the unresponsive casualty on his back, face up, and open his mouth. Kneel close to the side of the casualty's body.

1. Locate the lower edge of the casualty's ribs with your fingers. Run the fingers up along the rib cage to the notch (Figure 2-15A).

2. Place the middle finger on the notch and the index finger next to the middle finger on the lower edge of the breastbone. Place the heel of the other hand on the lower half of the breastbone next to the two fingers (Figure 2-15B).

3. Remove the fingers from the notch and place that hand on top of the positioned hand on the breastbone, extending or interlocking the fingers (Figure 2-15C).

4. Straighten and lock your elbows with your shoulders directly above your hands without bending the elbows, rocking, or allowing the shoulders to sag. Apply enough pressure to depress the breastbone 1 1/2 to 2 inches, then release the pressure completely (Figure 2-15D). Do this 6 to 10 times. Each thrust should be delivered quickly and distinctly. See Figure 2-16 for another view of the breastbone being depressed.

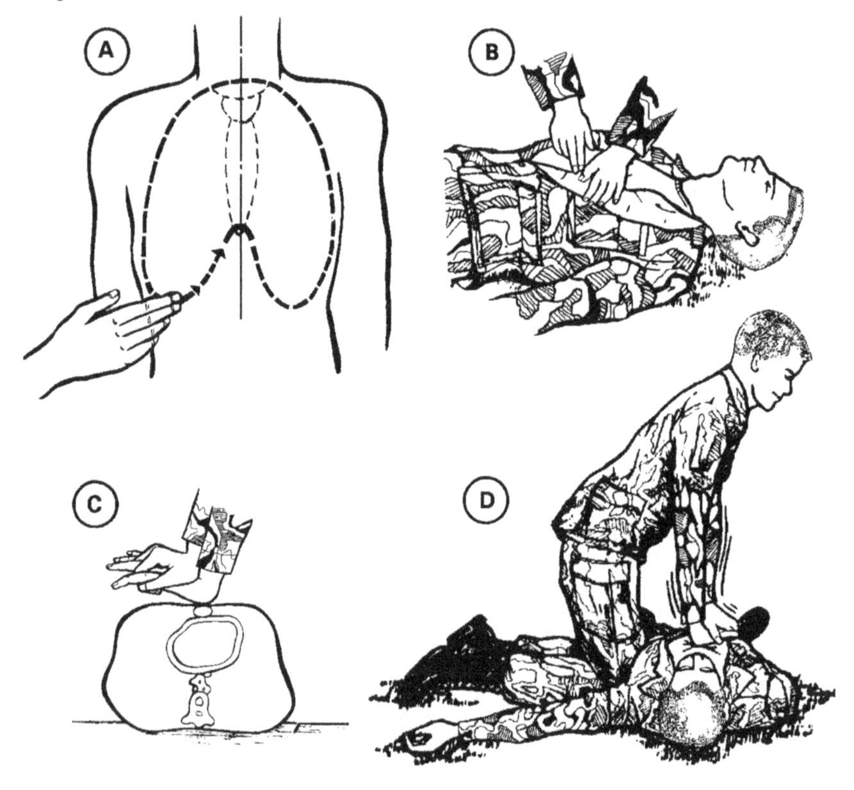

Figure 2-15. Hand placement for chest thrust (Illustrated A-D).

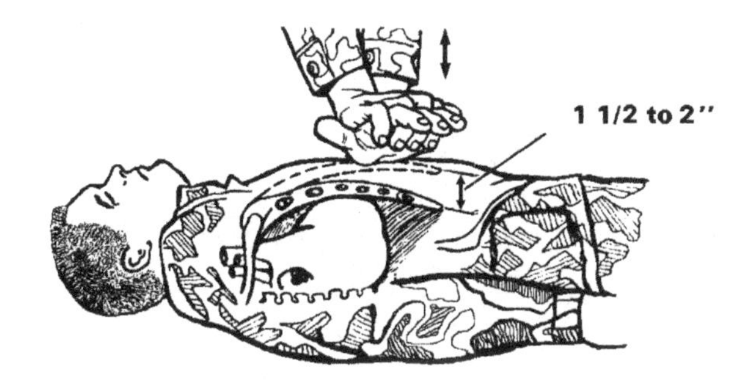

Figure 2-16. Breastbone depressed 1 1/2 to 2 inches.

(*b*) Repeat the sequence of chest thrust, finger sweep, and rescue breathing as long as necessary to clear the object from the obstructed airway. See paragraph (3) below.

(*c*) If the casualty's chest rises, proceed to feeling for his pulse.

(3) If you still cannot administer rescue breathing due to an airway obstruction, then remove the airway obstruction using the procedures in steps (*a*) and (*b*) below.

(*a*) Place the casualty on his back, face up, turn the unresponsive casualty as a unit, and call out for help.

(*b*) Perform finger sweep, keep casualty face up, use tongue-jaw lift to open mouth.

1. Open the casualty's mouth by grasping both his tongue and lower jaw between your thumb and fingers and lifting (tongue-jaw lift) (Figure 2-17). If you are unable to open his mouth, cross your fingers and thumb (crossed-finger method) and push his teeth apart (Figure 2-18) by pressing your thumb against his upper teeth and pressing your finger against his lower teeth.

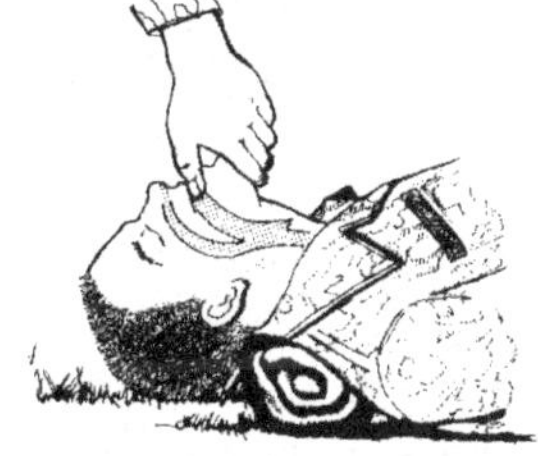

Figure 2-17. Opening casualty's mouth (tongue-jaw lift).

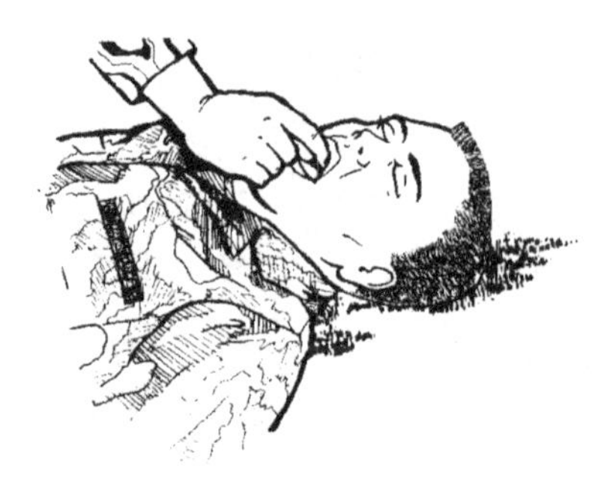

Figure 2-18. Opening casualty's mouth (crossed-finger method).

2. Insert the index finger of the other hand down along the inside of his cheek to the base of the tongue. Use a hooking motion from the side of the mouth toward the center to dislodge the foreign body (Figure 2-19).

Figure 2-19. Using finger to dislodge a foreign body.

WARNING

Take care not to force the object deeper into the airway by pushing it with the finger.

Section II. STOP THE BLEEDING AND PROTECT THE WOUND

2-13. General

The longer a service member bleeds from a major wound, the less likely he will be able to survive his injuries. It is, therefore, important that the first aid provider promptly stop the external bleeding.

2-14. Clothing

In evaluating the casualty for location, type, and size of the wound or injury, cut or tear his clothing and carefully expose the entire area of the wound. This procedure is necessary to properly visualize injury and avoid further contamination. Clothing stuck to the wound should be left in place to avoid further injury. DO NOT touch the wound; keep it as clean as possible.

WARNING

DO NOT REMOVE protective clothing in a chemical environment. Apply dressings over the protective clothing.

2-15. Entrance and Exit Wounds

Before applying the dressing, carefully examine the casualty to determine if there is more than one wound. A missile may have entered at one point and exited at another point. The *EXIT* wound is usually *LARGER* than the entrance wound.

WARNING

The casualty should be continually monitored for development of conditions which may require the performance of necessary basic lifesaving measures, such as clearing the airway and mouth-to-mouth resuscitation. All open (or penetrating) wounds should be checked for a point of entry and exit and first aid measures applied accordingly.

WARNING

If the missile lodges in the body (fails to exit), DO NOT attempt to remove it or probe the wound. Apply a dressing. If there is an object extending from (impaled in) the wound, DO NOT remove the object. Apply a dressing around the object and use additional improvised bulky materials/dressings (use the cleanest material available) to build up the area around the object to stabilize the object and prevent further injury. Apply a supporting bandage over the bulky materials to hold them in place.

2-16. Field Dressing

a. Use the casualty's field dressing; remove it from the wrapper and grasp the tails of the dressing with both hands (Figure 2-20).

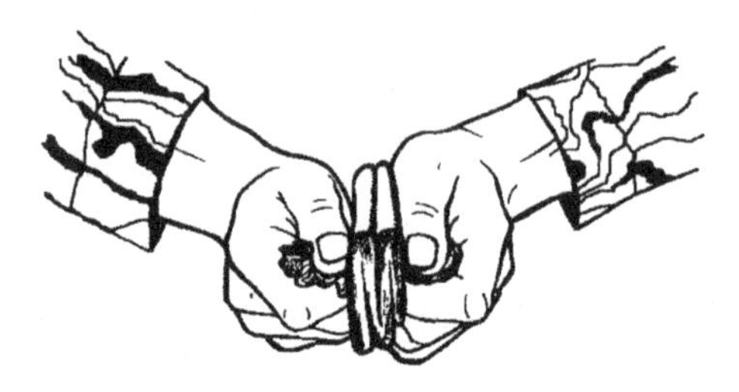

Figure 2-20. Grasping tails of dressing with both hands.

> **WARNING**
>
> **DO NOT touch the white (sterile) side of the dressing, and DO NOT allow it to come in contact with any surface other than the wound.**

b. Hold the dressing directly over the wound with the white side down. Pull the dressing open (Figure 2-21) and place it directly over the wound (Figure 2-22).

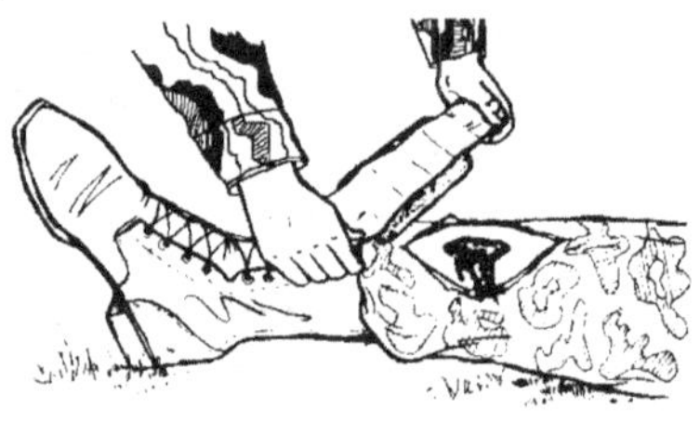

Figure 2-21. Pulling dressing open.

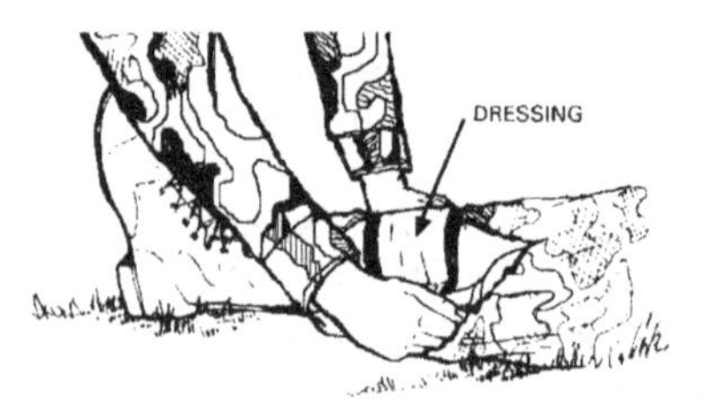

Figure 2-22. Placing dressing directly on wound.

c. Hold the dressing in place with one hand. Use the other hand to wrap one of the tails around the injured part, covering about one-half of the dressing (Figure 2-23). Leave enough of the tail for a knot. If the casualty is able, he may assist by holding the dressing in place.

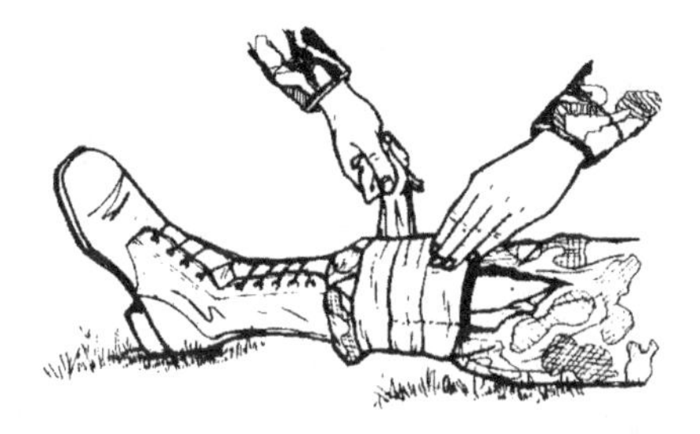

Figure 2-23. Wrapping tail of dressing around injured part.

d. Wrap the other tail in the opposite direction until the remainder of the dressing is covered. The tails should seal the sides of the dressing to keep foreign material from getting under it.

e. Tie the tails into a nonslip knot over the outer edge of the dressing (Figure 2-24). **DO NOT TIE THE KNOT OVER THE WOUND**. In order to allow blood to flow to the rest of an injured limb, tie the dressing firmly enough to prevent it from slipping but without causing a tourniquet-like effect; that is, the skin beyond the injury should not becomes cool, blue, or numb.

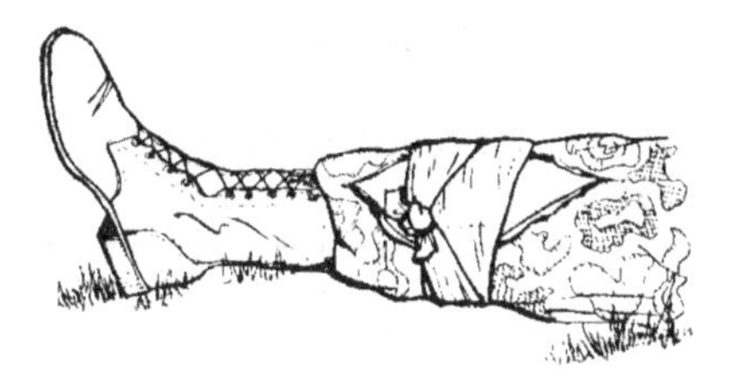

Figure 2-24. Tails tied into nonslip knot.

2-17. Manual Pressure

a. If bleeding continues after applying the sterile field dressing, direct manual pressure may be used to help control bleeding. Apply such pressure by placing a hand on the dressing and exerting firm pressure for 5 to 10 minutes (Figure 2-25). The casualty may be asked to do this himself if he is conscious and can follow instructions.

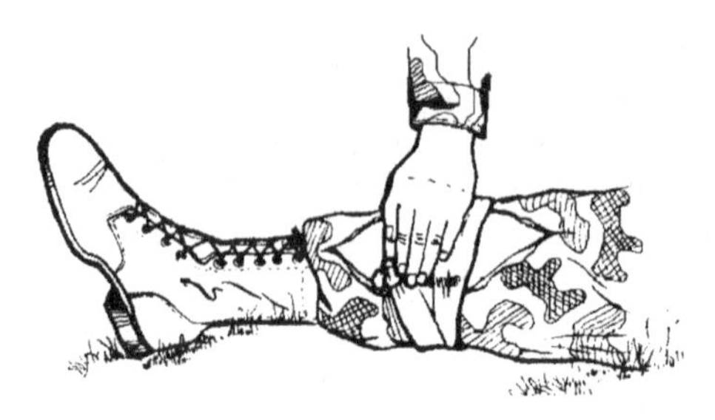

Figure 2-25. Direct manual pressure applied.

b. Elevate an injured limb slightly above the level of the heart to reduce the bleeding (Figure 2-26).

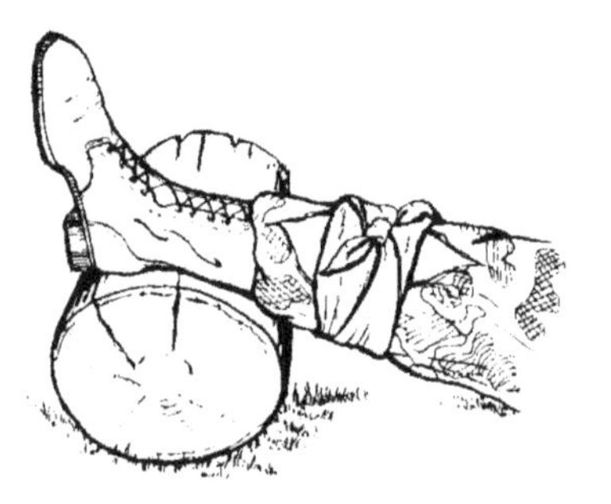

Figure 2-26. Injured limb elevated.

WARNING

DO NOT elevate a suspected fractured limb unless it has been properly splinted.

c. If the bleeding stops, check shock; administer first aid for shock as necessary. If the bleeding continues, apply a pressure dressing.

2-18. Pressure Dressing

Pressure dressings aid in blood clotting and compress the open blood vessel. If bleeding continues after the application of a field dressing, manual pressure, and elevation, then a pressure dressing must be applied as follows:

a. Place a wad of padding on top of the field dressing, directly over the wound (Figure 2-27). Keep the injured extremity elevated.

Figure 2-27. Wad of padding on top of field dressing.

NOTE

Improvised bandages may be made from strips of cloth. These strips may be made from T-shirts, socks, or other garments.

b. Place an improvised dressing (or cravat, if available) over the wad of padding (Figure 2-28). Wrap the ends tightly around the injured limb, covering the previously placed field dressing (Figure 2-29).

Figure 2-28. Improvised dressing over wad of padding

Figure 2-29. Ends of improvised dressing wrapped tightly around limb.

c. Tie the ends together in a nonslip knot, directly over the wound site (Figure 2-30). DO NOT tie so tightly that it has a tourniquet-like effect. If bleeding continues and all other measures have failed, or if the limb is severed, then apply a tourniquet. Use the tourniquet as a **LAST RESORT**. When the bleeding stops, check for shock; administer first aid for shock as necessary.

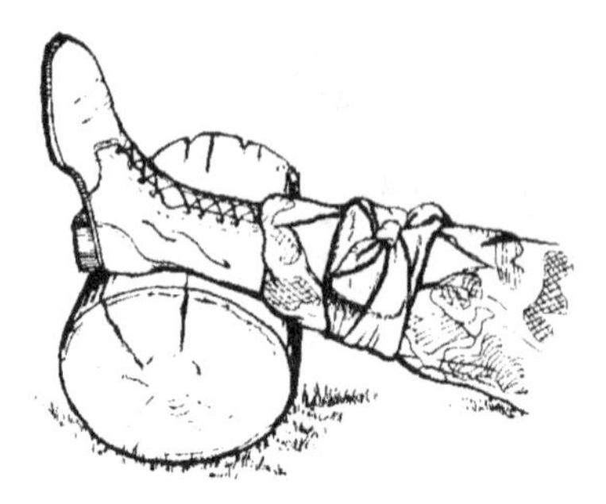

Figure 2-30. Ends of improvised dressing tied together in nonslip knot.

NOTE

Distal end of wounded extremities (fingers and toes) should be checked periodically for adequate circulation. The dressing must be loosened if the extremity becomes cool, blue, or numb.

NOTE

If bleeding continues and all other measures have failed (dressings and covering wound, applying direct manual pressure, elevating the limb above the heart level, and applying a pressure dressing while maintaining limb elevation) *then apply digital pressure* (see paragraph 2-19).

2-19. Digital Pressure

Digital pressure (often called "pressure points") is an alternative method to control bleeding. This method uses pressure from the fingers, thumbs, or hands to press at the site or point where a main artery supplying the wounded area lies near the skin surface or over bone (Figure 2-31). This pressure may help shut off or slow down the flow of blood from the heart to the wound and is used in combination with direct pressure and elevation. It may help in instances where bleeding is not easily controlled, where a pressure dressing has not yet been applied, or where pressure dressings are not readily available.

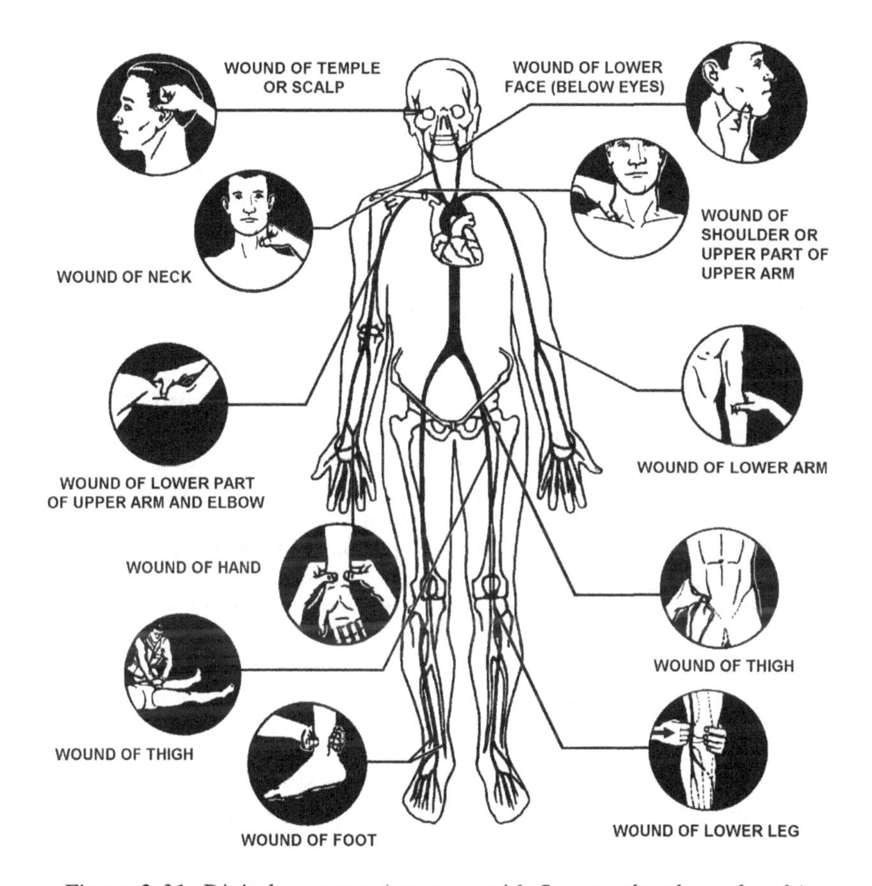

Figure 2-31. Digital pressure (pressure with fingers, thumbs or hands).

2-20. Tourniquet

DANGER

A tourniquet is only used on an arm or leg where there is a danger of the casualty losing his life (bleeding to death).

A tourniquet is a constricting band placed around an arm or leg to control bleeding. A service member whose arm or leg has been completely amputated may not be bleeding when first discovered, but a tourniquet should be applied anyway. This absence of bleeding is due to the body's normal defenses (contraction or clotting of blood vessels) as a result of the amputation, but after a period of time bleeding will start as the blood vessels relax or the clot may be knocked loose by moving the casualty. Bleeding from a major artery of the thigh, lower leg, or arm and bleeding from multiple arteries (which occurs in a traumatic amputation) may prove to be beyond control by manual pressure. If the pressure dressing (see paragraph 2-18, above) under firm hand pressure becomes soaked with blood and the wound continues to bleed, apply a tourniquet.

WARNING

Casualty should be continually monitored for development of conditions which may require the performance of necessary basic lifesaving measures, such as: clearing the airway, performing mouth-to-mouth resuscitation, preventing shock, and/or bleeding control. All open (or penetrating) wounds should be checked for a point of entry or exit and treated accordingly.

The tourniquet should not be used unless a pressure dressing has failed to stop the bleeding or an arm or leg has been cut off. On occasion, tourniquets have injured blood vessels and nerves. If left in place too long, a tourniquet can cause loss of an arm or leg. Once applied, it must stay in place, and the casualty must be taken to the nearest MTF as soon as possible. *DO NOT loosen or release a tourniquet after it has been applied as release could precipitate bleeding and potentially lead to shock.*

a. Improvising a Tourniquet. In the absence of a specially designed tourniquet, a tourniquet may be made from a strong, pliable material, such as gauze or muslin bandages, clothing, or cravats. An improvised tourniquet is used with a rigid stick-like object. To minimize skin damage, ensure that the improvised tourniquet is at least 2 inches wide.

WARNING

The tourniquet must be easily identified or easily seen.

WARNING

DO NOT use wire or shoestring for a tourniquet band.

b. Placing the Improvised Tourniquet.

(1) Place the tourniquet around the limb, between the wound and the body trunk (or between the wound and the heart). Never place it directly over a wound, a fracture, or joint. Tourniquets, for maximum effectiveness, should be placed on the upper arm or above the knee on the thigh (Figure 2-32).

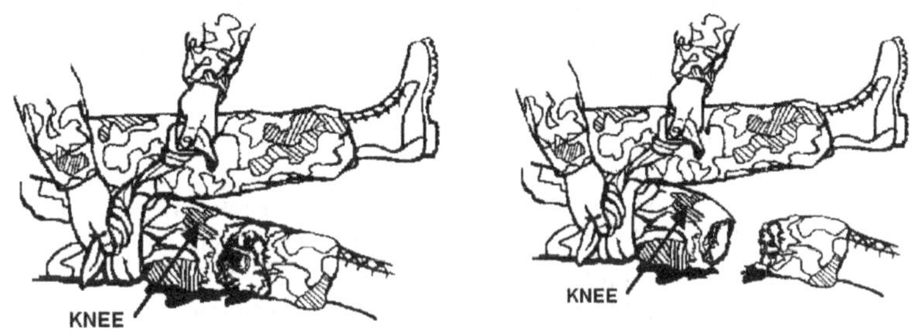

Figure 2-32. Tourniquet above knee.

(2) The tourniquet should be well-padded. If possible, place the tourniquet over the smoothed sleeve or trouser leg to prevent the skin from being pinched or twisted. If the tourniquet is long enough, wrap it around the limb several times, keeping the material as flat as possible. Damaging the skin may deprive the surgeon of skin required to cover an amputation. Protection of the skin also reduces pain.

c. Applying the Tourniquet.

(1) Tie a half-knot. (A half-knot is the same as the first part of tying a shoe lace.)

(2) Place a stick (or similar rigid object) on top of the half-knot (Figure 2-33).

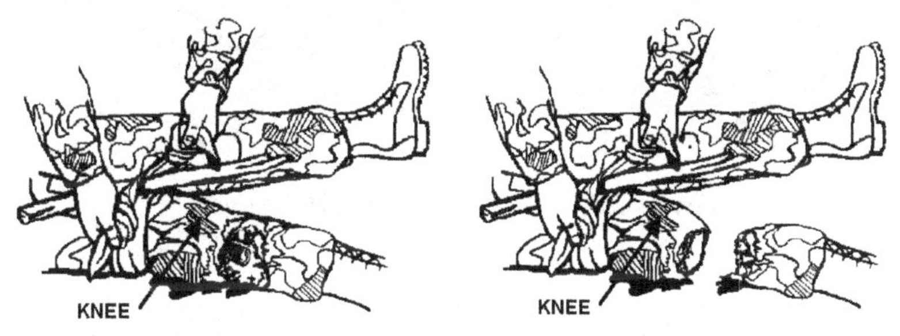

Figure 2-33. Rigid object on top of half-knot.

(3) Tie a full knot over the stick (Figure 2-34).

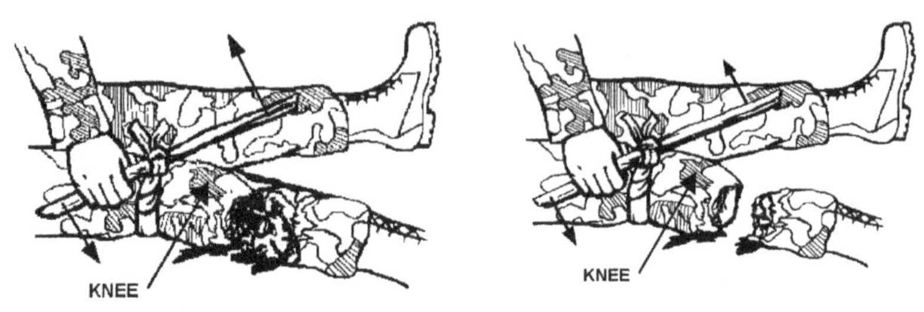

Figure 2-34. Full knot over rigid object.

(4) Twist the stick (Figure 2-35) until the tourniquet is tight around the limb and/or the bright red bleeding has stopped. In the case of amputation, dark oozing blood may continue for a short time. This is the blood trapped in the area between the wound and tourniquet.

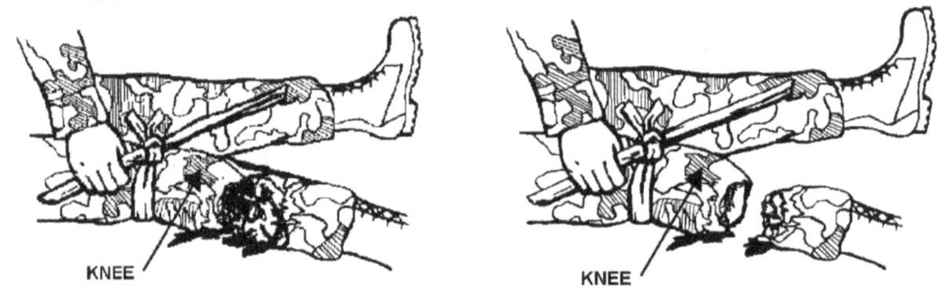

Figure 2-35. Stick twisted.

(5) Fasten the tourniquet to the limb by looping the free ends of the tourniquet over the ends of the stick. Then bring the ends around the limb to prevent the stick from loosening. Tie them together on the side of the limb (Figure 2-36).

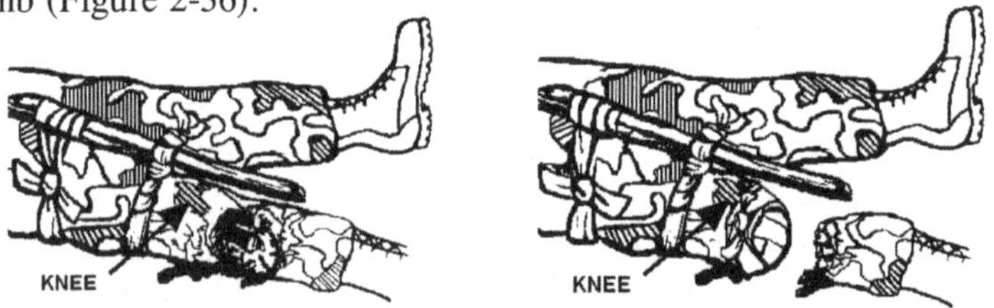

Figure 2-36. Tie free ends on side of limb.

NOTE

Other methods of securing the stick may be used as long as the stick does not unwind and no further injury results.

NOTE

If possible, save and transport any severed (amputated) limbs or body parts with (but out of sight of) the casualty.

(6) DO NOT cover the tourniquet—you should leave it in full view. If the limb is missing (total amputation), apply a dressing to the stump. All wounds should have a dressing to protect the wound from contamination.

(7) Mark the casualty's forehead with a "T" and the time to indicate a tourniquet has been applied. If necessary, use the casualty's blood to make this mark.

(8) Check and treat for shock.

(9) Seek medical aid.

CAUTION

Only appropriately skilled medical personnel may adjust or otherwise remove/release the tourniquet in the appropriate setting.

Section III. CHECK FOR SHOCK AND ADMINISTER FIRST AID MEASURES

2-21. General

The term *shock* has a variety of meanings. In medicine, it refers to a collapse of the body's cardiovascular system which includes an inadequate supply of blood to the body's tissues. Shock stuns and weakens the body. When the normal blood flow in the body is upset, death can result. Early recognition and proper first aid may save the casualty's life.

2-22. Causes and Effects

a. There are three basic mechanisms associated with shock. These are—

- The heart is damaged and fails to work as a pump.

- Blood loss (heavy bleeding) causes the volume of fluid within the vascular system to be insufficient.

- The blood vessels dilate (open wider) so that the blood within the system (even though it is a normal volume [the casualty is not bleeding or dehydrated]) is insufficient to provide adequate circulation within the body.

b. Shock may be the result of a number of conditions. These include—

- Dehydration.

- Allergic reaction to foods, drugs, insect stings, and snakebites.

- Significant loss of blood.

- Reaction to the sight of a wound, blood, or other traumatic scene.

- Traumatic injuries, such as—

  - Burns.

  - Gunshot or shrapnel wounds.

  - Crush injuries.

  - Blows to the body (which can cause broken bones or damage to internal organs).

  - Head injuries.

  - Penetrating wounds (such as from a knife, bayonet, or missile).

2-23. Signs and Symptoms of Shock

Examine the casualty to see if he has any of the following signs and symptoms:

- Sweaty but cool skin (clammy skin).

- Weak and rapid pulse.
- Paleness of skin (in dark-skinned individuals they may have a grayish look to their skin).
- Restlessness, nervousness.
- Thirst.
- Loss of blood (bleeding).
- Confusion (or loss of awareness).
- Faster-than-normal breathing rate.
- Blotchy or bluish skin (especially around the mouth and lips).
- Nausea and/or vomiting.

2-24. First Aid Measures for Shock

In the field, the first aid procedures administered for shock are identical to procedures that would be performed *to prevent shock.* When treating a casualty, assume that shock is present or will occur shortly. By waiting until actual signs and symptoms of shock are noticeable, the rescuer may jeopardize the casualty's life.

a. Position the Casualty. (DO NOT move the casualty or his limbs if suspected fractures have not been splinted. See Chapter 4 for details.)

(1) Move the casualty to cover, if cover is available and the situation permits.

(2) Lay the casualty on his back.

NOTE

A casualty in shock from a chest wound or one who is experiencing breathing difficulty, may breathe easier in a sitting position. If this is the case, allow him to sit upright, but monitor carefully in case his condition worsens.

(3) Elevate the casualty's feet higher than the level of his heart. Use a stable object (field pack or rolled up clothing) so that his feet will not slip off (Figure 2-37).

WARNING

DO NOT elevate legs if the casualty has an unsplinted broken leg, head injury, or abdominal injury.

Figure 2-37. Clothing loosened and feet elevated.

WARNING

Check casualty for leg fracture(s) and splint, if necessary, before elevating his feet. For a casualty with an abdominal wound, place his knees in an upright (flexed) position.

(4) Loosen clothing at the neck, waist, or wherever it may be binding.

CAUTION

DO NOT loosen or remove protective clothing in a chemical environment.

(5) Prevent chilling or overheating. The key is to maintain body temperature. In cold weather, place a blanket or other like item over him to keep him warm and under him to prevent chilling (Figure 2-38). However, if a tourniquet has been applied, leave it exposed (if possible). In hot weather, place the casualty in the shade and protect him from becoming chilled; however, avoid the excessive use of blankets or other coverings.

Figure 2-38. Body temperature maintained.

(6) Calm the casualty. Throughout the entire procedure of providing first aid for a casualty, the rescuer should reassure the casualty and keep him calm. This can be done by being authoritative (taking charge) and by showing self-confidence. Assure the casualty that you are there to help him.

(7) Seek medical aid.

b. Food and/or Drink. When providing first aid for shock, DO NOT give the casualty any food or drink. If you must leave the casualty or if he is unconscious, turn his head to the side to prevent him from choking if he vomits (Figure 2-39).

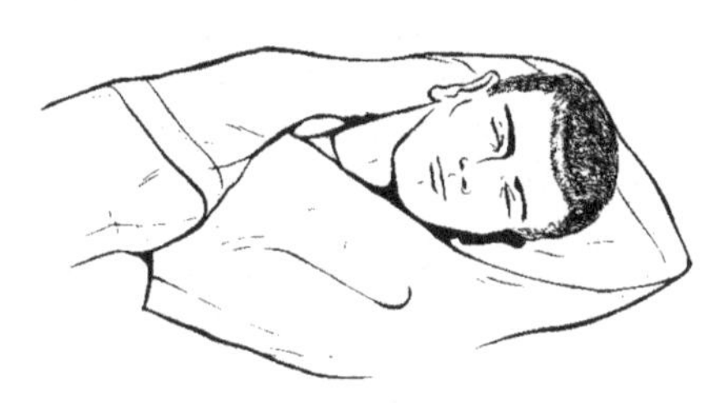

Figure 2-39. Casualty's head turned to side.

c. Evaluate Casualty. Continue to evaluate the casualty until medical personnel arrives or the casualty is transported to an MTF.

CHAPTER 3

FIRST AID FOR SPECIFIC INJURIES

3-1. General

Basic lifesaving steps are discussed in Chapters 1 and 2; they apply to first aid measures for all injuries. Some wounds and burns will require special precautions and procedures when applying these measures. This chapter discusses specific first aid procedures for wounds of the head, face, and neck; chest and stomach wounds; and burns. It also discusses the techniques for applying dressings and bandages to specific parts of the body.

3-2. Head, Neck, and Facial Injuries

a. Head Injuries.

(1) Head injuries range from minor abrasions or cuts on the scalp to severe brain injuries that may result in unconsciousness and sometimes death. Head injuries are classified as open or closed wounds. An open wound is one that is visible, has a break in the skin, and usually has evidence of bleeding. A closed wound may be visible (such as a depression in the skull) or the first aid provider may not be able to see any apparent injury (such as internal bleeding). Some head injuries result in unconsciousness; however, a service member may have a serious head wound and still be conscious. Casualties with head and neck injuries should be treated as though they also have a spinal injury. The casualty should not be moved until the head and neck is stabilized unless he is in immediate danger (such as close to a burning vehicle).

(2) Prompt first aid measures should be initiated for casualties with suspected head and neck injuries. The conscious casualty may be able to provide information on the extent of his injuries. However, as a result of the head injury, he may be confused and unable to provide accurate information. The signs and symptoms a first aid provider might observe are—

- Nausea and vomiting.
- Convulsions or twitches.
- Slurred speech.
- Confusion and loss of memory. (Does he know who he is? Does he know where he is? Does he know what day it is?)
- Recent unconsciousness.

- Dizziness.
- Drowsiness.
- Blurred vision, unequal pupils, or bruising (black eyes).
- Paralysis (partial or full).
- Complaint of headache.
- Bleeding or other fluid discharge from the scalp, nose, or ears.
- Deformity of the head (depression or swelling).
- Staggering while walking.

b. Neck Injuries. Neck injuries may result in heavy bleeding. Apply pressure above and below the injury, *but do not interfere with the breathing process*, and attempt to control the bleeding. Apply a dressing. Always evaluate the casualty for a possible neck fracture/spinal cord injury; if suspected, seek medical treatment immediately.

NOTE

Establish and maintain the airway in cases of facial or neck injuries. If a neck fracture or spinal cord injury is suspected, immobilize the injury and, if necessary, perform basic life support measures.

c. Facial Injuries. Soft tissue injuries of the face and scalp are common. Abrasions (scrapes) of the skin cause no serious problems. Contusions (injury without a break in the skin) usually cause swelling. A contusion of the scalp looks and feels like a lump. Laceration (cut) and avulsion (torn away tissue) injuries are also common. Avulsions are frequently caused when a sharp blow separates the scalp from the skull beneath it. Because the face and scalp are richly supplied with blood vessels (arteries and veins), wounds of these areas usually bleed heavily.

3-3. General First Aid Measures

a. General Considerations. The casualty with a head injury (or suspected head injury) should be continually monitored for the development of conditions that *may require* basic lifesaving measures. After initiating first aid measures, request medical assistance and evacuation. If dedicated medical evacuation assets are not available, transport the casualty to an MTF as soon as the situation permits. The first aid provider should not attempt to remove a protruding object from the head or give the casualty anything to eat or drink. Further, the first aid provider should be prepared to—

- Clear the airway.
- Control bleeding (external).
- Administer first aid measures for shock.
- Keep the casualty warm.
- Protect the wound.

b. Unconscious Casualty. An unconscious casualty does not have control of all of his body's functions and may choke on his tongue, blood, vomitus, or other substances. (Refer to Figure 2-39.)

(1) *Breathing*. The brain requires a constant supply of oxygen. A bluish (or in an individual with dark skin—grayish) color of skin around the lips and nail beds indicates that the casualty is not receiving enough oxygen. Immediate action must be taken to clear the airway, to position the casualty on his side, or to initiate rescue breathing.

(2) *Bleeding*. Bleeding from a head injury usually comes from blood vessels within the scalp. Bleeding can also develop inside the skull or within the brain. In most instances visible bleeding from the head can be controlled by application of the field first aid dressing.

CAUTION

DO NOT attempt to put unnecessary pressure on the wound or attempt to push any brain matter back into the head (skull). **DO NOT** apply a pressure dressing.

c. Concussion. If an individual receives a heavy blow to the head or face, he may suffer a brain concussion (an injury to the brain that involves a temporary loss of some or all of the brain's ability to function). For example, the casualty may not breathe properly for a short period of time, or he may become confused and stagger when he attempts to walk. Symptoms of a concussion may only last for a short period of time. However, if a casualty is suspected of having suffered a concussion, he should be transported to an MTF as soon as conditions permit.

d. Convulsions. Convulsions (seizures/involuntary jerking) may occur even after a mild head injury. When a casualty is convulsing, protect him from hurting himself. Take the following measures:

(1) Ease him to the ground if he is standing or sitting.

(2) Support his head and neck.

(3) Maintain his airway.

(4) Protect him from further injury (such as hitting close-by objects).

NOTE

DO NOT forcefully hold the arms and legs if they are jerking because this can lead to broken bones. **DO NOT** force anything between the casualty's teeth—especially if they are tightly clenched because this may obstruct the casualty's airway. Maintain the casualty's airway if necessary.

e. Brain Damage. In *severe* head injuries where brain tissue is protruding, *leave the wound alone*; carefully place a loose moistened dressing (moistened with sterile normal saline if available) and also a first aid dressing over the tissue to protect it from further contamination. ***DO NOT*** *remove or disturb any foreign matter that may be in the wound*. Position the casualty so that his head is higher than his body. Keep him warm and *seek medical assistance immediately*.

NOTE

If there is an object extending from the wound, **DO NOT** remove the object. Improvise bulky dressings from the cleanest material available and place this material around the protruding object for support, then apply the field dressing.

3-4. Chest Wounds

Blunt trauma, bullet or missile wounds, stab wounds, or falls may cause chest injuries. These injuries can be serious and may cause death quickly if first aid is not administered in a timely manner. A casualty with a chest injury may complain of pain in the chest or shoulder area; he may have difficulty breathing. His chest may not rise normally when he breathes. The injury may cause the casualty to cough up blood and to have a rapid or a weak heartbeat. A casualty with an open chest wound has a punctured chest wall. The sucking sound heard when he breathes is caused by air leaking into his chest cavity. This particular type of wound is dangerous and will collapse the injured lung (Figure 3-1). Breathing becomes difficult for the casualty because the wound is open. The service members life may depend upon how quickly you apply an occlusive dressing over the wound (refer to paragraph 3-5).

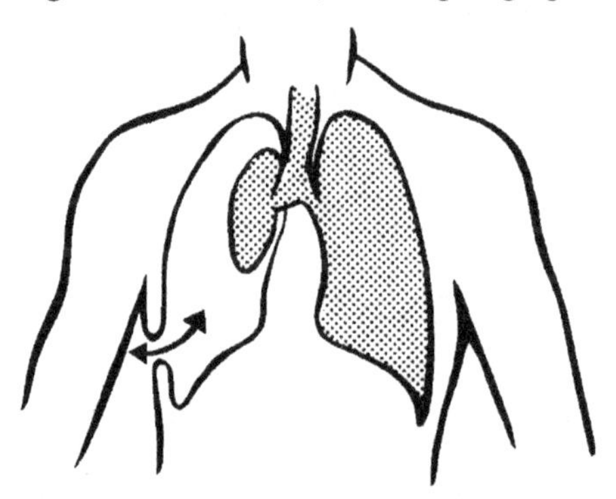

Figure 3-1. Collapsed lung.

3-5. First Aid for Chest Wounds

a. Evaluate the Casualty. Be prepared to perform first aid measures. These measures may include clearing the airway, rescue breathing, treatment for shock, and/or bleeding control.

b. Expose the Wound. If appropriate, cut or remove the casualty's clothing to expose the wound. Remember, **DO NOT** remove clothing that is stuck to the wound because additional injury may result. **DO NOT** attempt to clean the wound.

NOTE

Examine the casualty to see if there is an entry and exit wound. If there are two wounds (entry, exit), perform the same procedure for both wounds. Treat the more serious (heavier bleeding, larger) wound first. It may be necessary to improvise a dressing for the second wound by using strips of cloth, such as a torn T-shirt, or whatever material is available. Also, listen for sucking sounds to determine if the chest wall is punctured.

CAUTION

If there is an object impaled in the wound, **DO NOT** remove it. Apply a dressing around the object and use additional improvised bulky materials/dressings (use the cleanest materials available) to build up the area around the object. Apply a supporting bandage over the bulky materials to hold them in place.

CAUTION

DO NOT REMOVE protective clothing in a chemical environment. Apply dressings *over* the protective clothing.

c. Open the Casualty's Field Dressing Plastic Wrapper. In cases where there is a sucking chest wound, the plastic wrapper is used with the field dressing to create an occlusive dressing. If a plastic wrapper is not available, or if an additional wound needs to be treated; cellophane, foil, the casualty's poncho, or similar material may be used. The covering should be wide enough to extend 2 inches or more beyond the edges of the wound in all directions.

(1) Tear open one end of the casualty's plastic wrapper covering the field dressing. Be careful not to destroy the wrapper and **DO NOT** touch the inside of the wrapper.

(2) Remove the inner packet (field dressing).

(3) Complete tearing open the empty plastic wrapper using as much of the wrapper as possible to create a flat surface.

d. Place the Wrapper Over the Wound. Place the inside surface of the plastic wrapper directly over the wound *when the casualty exhales* and hold it in place (Figure 3-2). The casualty may hold the plastic wrapper in place if he is able.

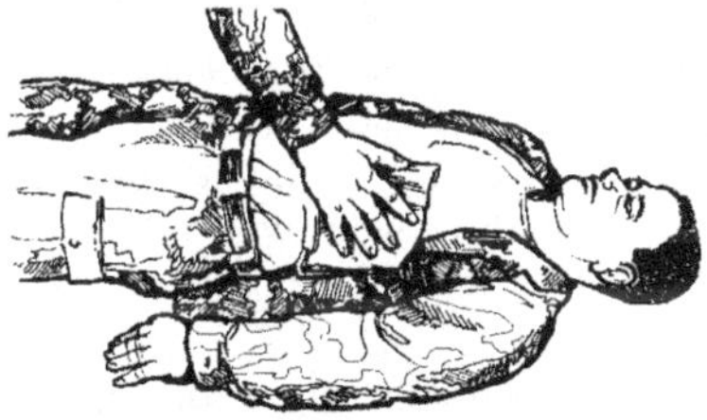

Figure 3-2. Open chest wound sealed with an occlusive dressing.

e. Apply the Dressing to the Wound.

(1) Use your free hand and shake open the field dressing (Figure 3-3).

Figure 3-3. Shaking open the field dressing.

(2) Place the white side of the dressing on the plastic wrapper covering the wound (Figure 3-4).

Figure 3-4. Field dressing placed on plastic wrapper.

NOTE

Use the casualty's field dressing, not your own.

(3) Have the casualty breathe normally.

(4) While maintaining pressure on the dressing, grasp one tail of the field dressing with the other hand and wrap it around the casualty's back. If tape is available, tape three sides of the plastic wrapper to the chest wall to provide occlusive type dressing. Leave one side untapped to provide emergency escape for air that may build up in the chest. If tape is not available, secure wrapper on three sides with field dressing leaving the fourth side as a flap.

(5) Wrap the other tail in the opposite direction, bringing both tails over the dressing (Figure 3-5).

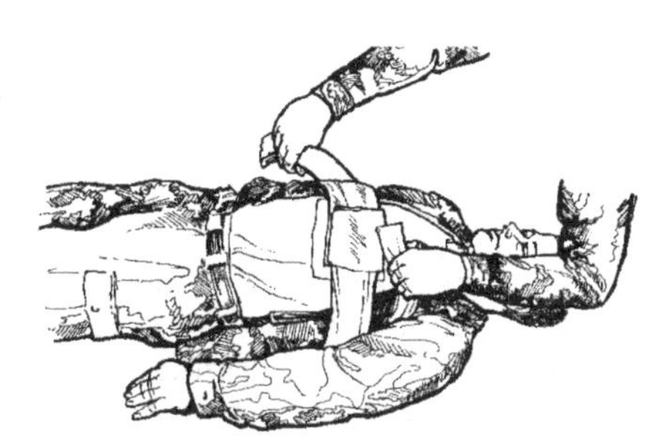

Figure 3-5. Tails of field dressing wrapped around casualty in opposite direction.

(6) Tie the tails into a square knot in the center of the dressing *after* the casualty exhales and *before* he inhales. This will aid in maintaining pressure on the bandage after it has been tied (Figure 3-6). Tie the dressing firmly enough to secure the dressing without interfering with the casualty's breathing.

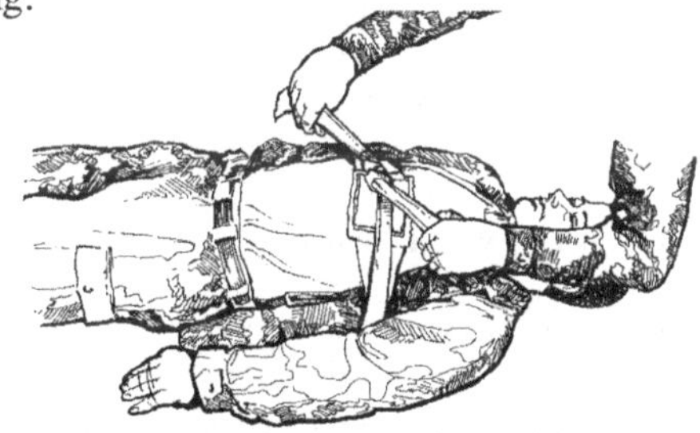

Figure 3-6. Tails of dressing tied into square knot over center of dressing.

NOTE

When practical, apply direct manual pressure over the dressing for 5 to 10 minutes to help control the bleeding.

f. *Position the Casualty*. Position the casualty on his injured side or in a sitting position, whichever makes breathing easier (Figure 3-7).

Figure 3-7. Casualty positioned (lying) on injured side.

g. Seek Medical Assistance. Contact medical personnel.

WARNING

If an occlusive dressing has been improperly placed, air may enter the chest cavity with no means of escape. This causes a life-threatening condition called *tension pneumothorax*. If the casualty's condition (for example, difficulty breathing, shortness of breath, restlessness, or blueness/grayness of the skin) worsens after placing the dressing, quickly lift or remove, and then replace the occlusive dressing.

3-6. Abdominal Wounds

The most serious abdominal wound is one in which an object penetrates the abdominal wall and pierces internal organs or large blood vessels. In these instances, bleeding may be severe and death can occur rapidly.

3-7. First Aid for Abdominal Wounds

a. Evaluate the Casualty. Be prepared to perform basic first aid measures. Always check for both entry and exit wounds. If there are two wounds (entry and exit), treat the wound that appears more serious first (for example, the heavier bleeding, protruding organs, larger wound, and so forth). It may be necessary to improvise dressings for the second wound by using strips of cloth, a T-shirt, or the cleanest material available.

b. Position the Casualty. Place and maintain the casualty on his back with his knees in an upright (flexed) position (Figure 3-8). The knees-up position helps relieve pain, assists in the treatment of shock, prevents further exposure of the bowel (intestines) or abdominal organs, and helps relieve abdominal pressure by allowing the abdominal muscles to relax.

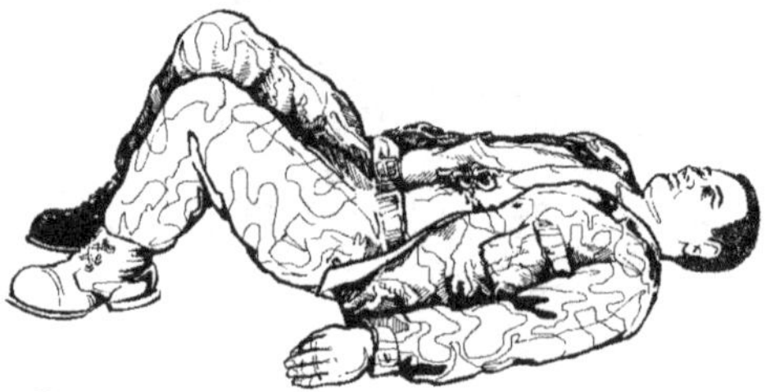

Figure 3-8. Casualty positioned (lying) on back with knees (flexed) up.

c. Expose the Wound.

(1) Remove the casualty's loose clothing to expose the wound. However, **DO NOT** attempt to remove clothing that is stuck to the wound; removing it may cause further injury.

CAUTION

DO NOT REMOVE protective clothing in a chemical environment. Apply dressings *over* the protective clothing.

(2) Gently pick up any organs that may be on the ground. Do this with a clean, dry dressing or with the cleanest available material. Place the organs on top of the casualty's abdomen (Figure 3-9).

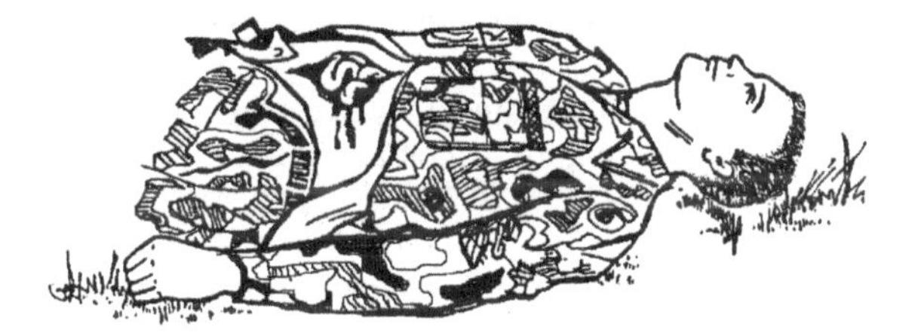

Figure 3-9. Protruding organs placed near wound.

NOTE

DO NOT probe, clean, or try to remove any foreign object from the abdomen. **DO NOT** touch with bare hands any exposed organs. **DO NOT** push organs back inside the body.

d. Apply the Field Dressing. Use the casualty's field dressing, not your own. If the field dressing is not large enough to cover the entire wound, the plastic wrapper from the dressing may be used to cover the wound first (placing the field dressing on top). Open the plastic wrapper carefully without touching the inner surface. If necessary, other improvised dressings may be made from clothing, blankets, or the cleanest materials available.

WARNING

If there is an object extending from the wound, DO NOT remove it. Place as much of the wrapper over the wound as possible without dislodging or moving the object. DO NOT place the wrapper over the object.

(1) Grasp the tails in both hands.

(2) Hold the dressing with the white side down directly over the wound. **DO NOT** touch the white (sterile) side of the dressing or allow anything except the wound to come in contact with it.

(3) Pull the dressing open and place it directly over the wound (Figure 3-10). If the casualty is able, he may hold the dressing in place.

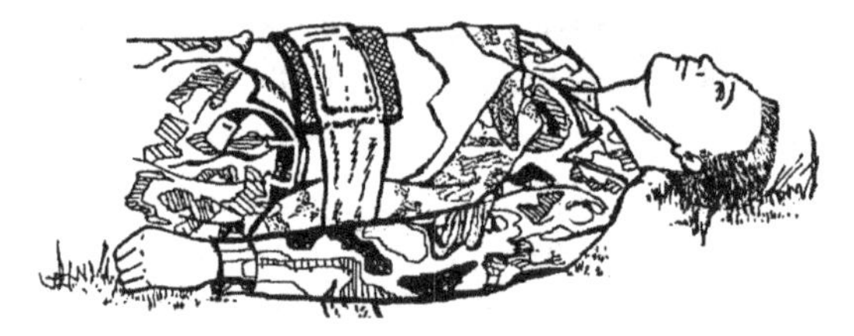

Figure 3-10. Dressing placed directly over the wound.

(4) Hold the dressing in place with one hand and use the other hand to wrap one of the tails around the body.

(5) Wrap the other tail in the opposite direction until the dressing is completely covered. Leave enough of the tail for a knot.

(6) Loosely tie the tails with a square knot at the casualty's side (Figure 3-11).

Figure 3-11. Dressing applied and tails tied with a square knot.

WARNING

When the dressing is applied, DO NOT put pressure on the wound or exposed internal parts, because pressure could cause further injury (vomiting, ruptured intestines, and so forth). Therefore, tie the dressing ties (tails) loosely at casualty's side, not directly over the dressing.

(7) Tie the dressing firmly enough to prevent slipping without applying pressure to the wound site (Figure 3-12).

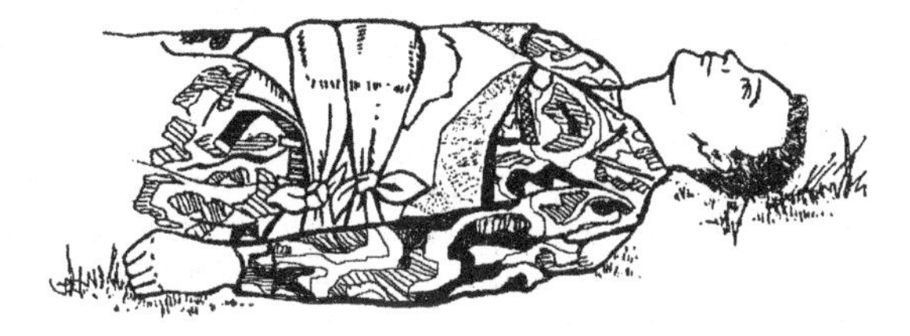

Figure 3-12. Field dressing covered with improvised material and loosely tied.

Field dressings can be covered with improvised reinforcement material (cravats, strips of torn T-shirt, or other cloth) for additional support and protection. Tie improvised bandage on the opposite side of the dressing ties firmly enough to prevent slipping but without applying additional pressure to the wound.

CAUTION

DO NOT give casualties with abdominal wounds food or water (moistening the lips is allowed).

e. Seek Medical Assistance. Notify medical personnel.

3-8. Burn Injuries

Burns often cause extreme pain, scarring, or even death. Before administering first aid, you must be able to recognize the type of burn. There are four types of burns:

- Thermal burns caused by fire, hot objects, hot liquids, and gases; or by nuclear blast or fireball.
- Electrical burns caused by electrical wires, current, or lightning.
- Chemical burns caused by contact with wet or dry chemicals or white phosphorus (WP)—from marking rounds and grenades.
- Laser burns (eye [ocular] injury).

3-9. First Aid for Burns

a. Eliminate the Source of the Burn. The source of the burn must be eliminated before any evaluation of the casualty can occur and first aid administered.

(1) Quickly remove the casualty from danger and cover the *thermal burn* with any large nonsynthetic material, such as a field jacket. If the casualty's clothing is still on fire, roll the casualty on the ground to smother (put out) the flames (Figure 3-13).

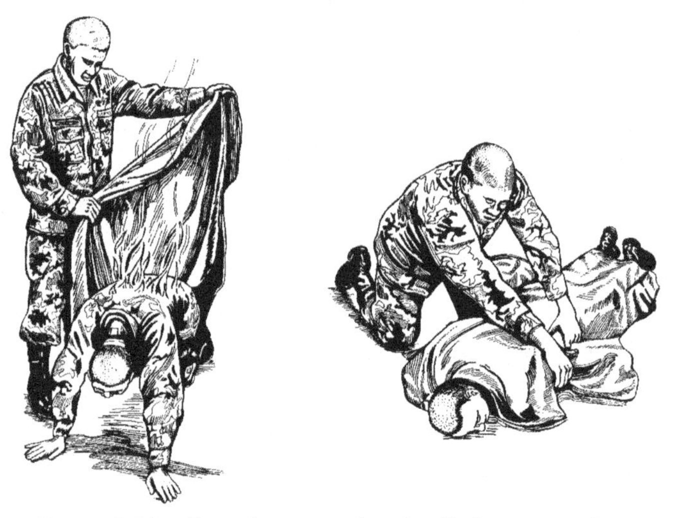

Figure 3-13. Casualty covered and rolled on ground.

> **CAUTION**
>
> Synthetic materials, such as nylon, may melt and cause further injury.

(2) Remove the *electrical burn* casualty from the electrical source by turning off the electrical current. **DO NOT** attempt to turn off the electricity if the source is not close by. Speed is critical, so **DO NOT** waste unnecessary time. If the electricity cannot be turned off, wrap any *nonconductive* material (*dry* rope, clothing, wood, and so forth) around the casualty's back and shoulders and drag the casualty away from the electrical source (Figure 3-14). **DO NOT** make body-to-body contact with the casualty or touch any wires because you could also become an electrical burn casualty.

Figure 3-14. Casualty removed from electrical source (using nonconductive material).

WARNING

High voltage electrical burns may cause temporary unconsciousness, difficulties in breathing, or difficulties with the heart (heartbeat).

(3) Remove the *chemical* from the *burned casualty*. Remove *liquid* chemicals by flushing with as much water as possible. Remove *dry* chemicals by brushing off loose particles (**DO NOT** use the bare surface of your hand because you could become a chemical burn casualty) and then flush with large amounts of water, if available. If large amounts of water are not available, then **NO** water should be applied because small amounts of water applied to a dry chemical burn may cause a chemical reaction. When WP strikes the skin, smother with a wet cloth or mud. Keep WP covered with a wet material to exclude air; this should help prevent the particles from burning.

(4) Remove the *laser burn* casualty from the source. When removing the casualty from the laser beam source, be careful not to enter the

beam or you may become a casualty. Never look directly at the beam source and if possible, wear appropriate eye protection.

NOTE

After the casualty is removed from the source of the burn, he should be evaluated for conditions requiring basic first aid measures.

b. Expose the Burn. Cut and gently lift away any clothing covering the burned area, without pulling clothing over the burns. Leave in place any clothing that is stuck to the burn. If the casualty's hands or wrists have been burned, remove jewelry if possible without causing further injury (rings, watches, and so forth) and place in his pockets. This prevents the necessity to cut off jewelry since swelling usually occurs as a result of a burn.

CAUTION

DO NOT lift or cut away clothing if in a chemical environment. Apply the dressing directly over the casualty's protective clothing. **DO NOT** attempt to decontaminate skin where blisters have formed.

c. Apply a Field Dressing to the Burn.

(1) Grasp the tails of the casualty's dressing in both hands.

(2) Hold the dressing directly over the wound with the white side down, pull the dressing open, and place it directly over the wound. **DO NOT** touch the white (sterile) side of the dressing or allow anything except the wound to come in contact with it. If the casualty is able, he may hold the dressing in place.

(3) Hold the dressing in place with one hand and use the other hand to wrap one of the tails around the limbs or the body.

(4) Wrap the other tail in the opposite direction until the dressing is completely covered.

(5) Tie the tails into a square knot over the outer edge of the dressing. The dressing should be applied lightly over the burn. Ensure that dressing is applied firmly enough to prevent it from slipping.

NOTE

Use the cleanest improvised dressing material available if a field dressing is not available or if it is not large enough for the entire wound.

d. *Take the Following Precautions*:

- **DO NOT** place the dressing over the face or genital area.
- **DO NOT** break the blisters.
- **DO NOT** apply grease or ointments to the burns.
- For electrical burns, check for both an entry and exit burn from the passage of electricity through the body. Exit burns may appear on any area of the body despite location of entry burn.
- For burns caused by wet or dry chemicals, flush the burns with large amounts of water and cover with a dry dressing.
- For burns caused by WP, flush the area with water, then cover with a wet material, dressing, or mud to exclude the air and keep the WP particles from burning.
- For laser burns, apply a field dressing.
- If the casualty is conscious and not nauseated, give him small amounts of water.

e. *Seek Medical Assistance*. Notify medical personnel.

3-10. Dressings and Bandages

a. *Head Wounds.*

(1) *Position the casualty.*

WARNING

DO NOT move the casualty if you suspect he has sustained a neck, spine, or head injury (which produces any signs or symptoms other than minor bleeding).

- If the casualty has a minor (superficial) scalp wound and is conscious:

  - Have the casualty sit up (unless other injuries prohibit or he is unable to).

  - If the casualty is lying down and is not accumulating fluids or drainage in his throat, elevate his head slightly.

  - If the casualty is bleeding from or into his mouth or throat, turn his head to the side or position him on his side so that the airway will be clear. Avoid putting pressure on the wound and place him on his uninjured side (Figure 3-15).

Figure 3-15. Casualty lying on side opposite injury.

- If the casualty is unconscious or has a severe head injury, then suspect and treat him as having a potential neck or spinal injury, *immobilize and **DO NOT** move the casualty.*

NOTE

If the casualty is choking or vomiting or is bleeding from or into his mouth (thus compromising his airway), position him on his uninjured side to allow for drainage and to help keep his airway clear.

WARNING

If it is necessary to turn a casualty with a suspected neck/spine injury; roll the casualty gently onto his side, keeping the head, neck, and body aligned while providing support for the head and neck. DO NOT roll the casualty by yourself but seek assistance. *Move him only if absolutely necessary*, otherwise keep the casualty immobilized to prevent further damage to the neck/spine.

(2) *Expose the wound.* Remove the casualty's helmet (if necessary). In a nuclear, biological, and chemical (NBC) environment, the first aid provider must leave the casualty as much protection (such as protective mask, mission-oriented protective posture [MOPP] overgarments) as possible. What items of protective equipment can be removed is dependent upon the casualty's injuries (where on the body and what type), the MOPP level, integrity of protective equipment (such as tears in the garment or mask seal), availability of chemical protective shelters, and the tactical situation.

WARNING

DO NOT attempt to clean the wound or remove a protruding object.

NOTE

Always use the casualty's field dressing, not your own.

(3) *Apply a dressing to a wound of the forehead or back of head.* To apply a dressing to a wound of the forehead or back of the head—

(*a*) Remove the dressing from the wrapper.

(*b*) Grasp the tails of the dressing in both hands.

(*c*) Hold the dressing (white side down) directly over the wound. **DO NOT** touch the white (sterile) side of the dressing or allow anything except the wound to come in contact with it.

(*d*) Place it directly over the wound.

(*e*) Hold it in place with one hand. If the casualty is able, he may assist.

(*f*) Wrap the first tail horizontally around the head; ensure the tail covers the dressing (Figure 3-16).

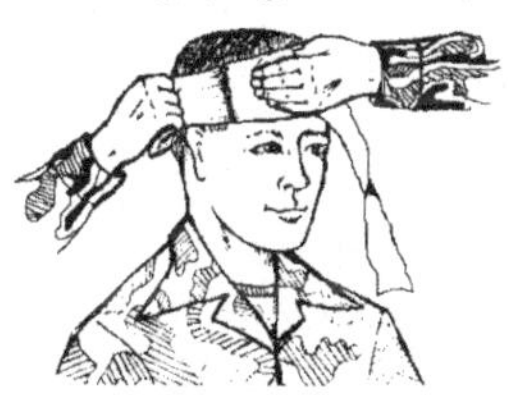

Figure 3-16. First tail of dressing wrapped horizontally around head.

(*g*) Hold the first tail in place and wrap the second tail in the opposite direction, covering the dressing (Figure 3-17).

Figure 3-17. Second tail wrapped in opposite direction.

(*h*) Tie a square knot and secure the tails at the side of the head, making sure they **DO NOT** cover the eyes or ears (Figure 3-18).

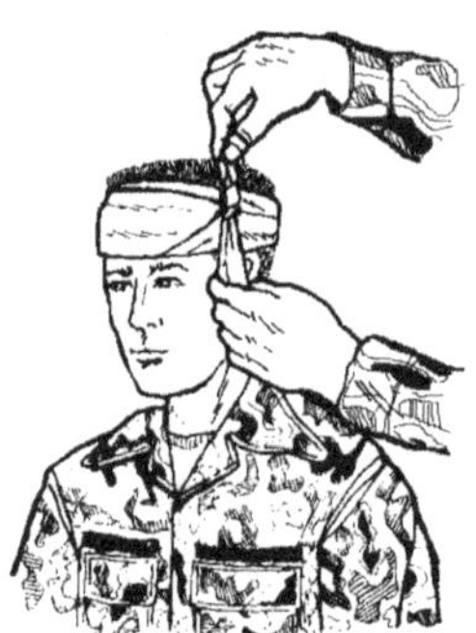

Figure 3-18. Tails tied in square knot at side of head.

(4) *Apply a dressing to a wound on top of the head.* To apply a dressing to a wound on top of the head—

(*a*) Remove the dressing from the wrapper.

(*b*) Grasp the tails of the dressing in both hands.

(*c*) Hold it (white side down) directly over the wound. **DO NOT** touch the white (sterile) side of the dressing or allow anything except the wound to come in contact with it.

(*d*) Place it over the wound (Figure 3-19).

Figure 3-19. Dressing placed over wound.

(*e*) Hold it in place with one hand. If the casualty is able, he may assist.

(*f*) Wrap one tail down under the chin (Figure 3-20), up in front of the ear, over the dressing, and in front of the other ear.

Figure 3-20. One tail of dressing wrapped under chin.

WARNING

Ensure the tails remain wide and close to the front of the chin to avoid choking the casualty.

(*g*) Wrap the remaining tail under the chin in the opposite direction and up the side of the face to meet the first tail (Figure 3-21).

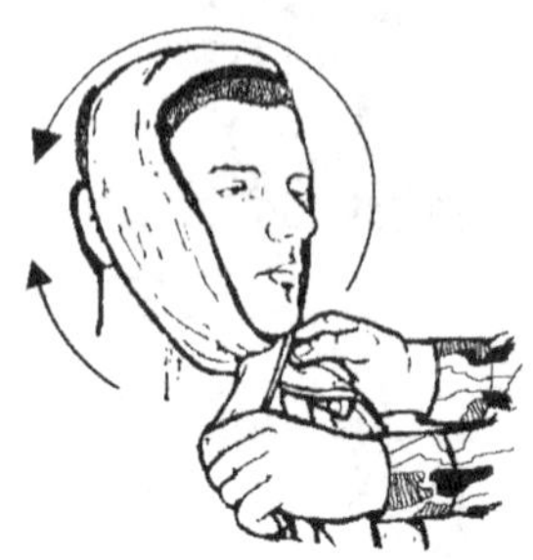

Figure 3-21. Remaining tail wrapped under chin in opposite direction.

(*h*) Cross the tails (Figure 3-22), bringing one around the forehead (above the eyebrows) and the other around the back of the head (at the base of the skull) to a point just above and in front of the opposite ear, and tie them using a square knot (Figure 3-23).

Figure 3-22. Tails of dressing crossed with one around forehead.

Figure 3-23. Tails tied in square knot (in front of and above ear).

(5) *Apply a triangular bandage to the head.* To apply a triangular bandage to the head—

(*a*) Turn the base (longest side) of the bandage up and center its base on the center of the forehead, letting the point (apex) fall on the back of the neck (Figure 3-24A).

(*b*) Take the ends behind the head and cross the ends over the apex.

(*c*) Take them over the forehead and tie them (Figure 3-24B).

(*d*) Tuck the apex behind the crossed part of the bandage or secure it with a safety pin, if available (Figure 3-24C).

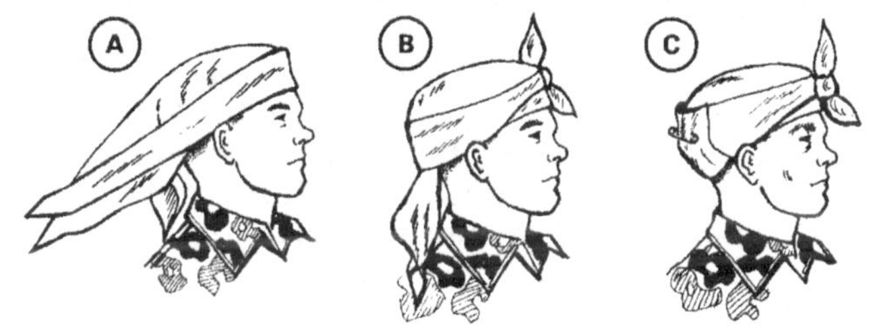

Figure 3-24. Triangular bandage applied to head (Illustrated A—C)

(6) *Apply a cravat bandage to the head.* To apply a cravat bandage to the head—

(*a*) Place the middle of the bandage over the dressing (Figure 3-25A).

(*b*) Cross the two ends of the bandage in opposite directions completely around the head (Figure 3-25B).

(*c*) Tie the ends over the dressing (Figure 3-25C).

Figure 3-25. Cravat bandage applied to head (Illustrated A—C).

b. Eye Injuries. The eye is a vital sensory organ, and blindness is a severe physical handicap. Timely first aid of the eye may relieve pain and may also help to prevent shock, permanent eye injury, and possible loss of vision. Because the eye is very sensitive, any injury can be easily aggravated if it is improperly handled. Injuries of the eye may be quite severe. Cuts of the eyelids can appear to be very serious, but if the eyeball is not involved, a person's vision usually will not be damaged. However, lacerations (cuts) of the eyeball can cause permanent damage or loss of sight.

(1) *Lacerated/torn eyelids*. Lacerated eyelids may bleed heavily, but bleeding usually stops quickly. Cover the injured eye with a sterile dressing. **DO NOT** put pressure on the wound because you may injure the eyeball. Handle torn eyelids very carefully to prevent further injury. Place any detached pieces of the eyelid on a clean bandage or dressing and immediately send them with the casualty to the medical facility.

(2) *Lacerated eyeball (injury to the globe)*. Lacerations or cuts to the eyeball may cause serious and permanent eye damage. Cover the injury with a loose sterile dressing. **DO NOT** put pressure on the eyeball because additional damage may occur. An important point to remember is that when one eyeball is injured, you should immobilize both eyes. This is done by applying a bandage to both eyes. Because the eyes move together, covering both will lessen the chances of further damage to the injured eye. (However, in hazardous surroundings, leave uninjured eye uncovered to enable casualty to see.)

CAUTION

DO NOT apply pressure when there is a possible laceration of the eyeball. The eyeball contains fluid. Pressure applied over the eye will force the fluid out, resulting in permanent injury. **APPLY PROTECTIVE DRESSING WITHOUT ADDED PRESSURE**.

(3) *Extruded eyeballs*. Service members may encounter casualties with severe eye injuries that include an extruded eyeball (eyeball out-of-socket). In such instances you should gently cover the extruded eye with a loose moistened dressing and also cover the unaffected eye. **DO NOT** bind or exert pressure on the injured eye while applying the dressing. Keep the casualty quiet, place him on his back, treat for shock, and evacuate him immediately.

(4) *Burns of the eyes*. Chemical burns, thermal (heat) burns, and light burns can affect the eyes.

(*a*) *Chemical burns*. Injuries from chemical burns require immediate first aid. Mainly acids or alkalies cause chemical burns. The first aid measures consist of flushing the eyes immediately with large amounts of water for at least 5 to 20 minutes, or as long as necessary to flush out the chemical and, once flushed, bandaging the eyes. If the burn is an acid burn, you should flush the eye for at least 5 to 10 minutes. If the burn is an alkali burn, you should flush the eye for at least 20 minutes. After the eye has been flushed evacuate the casualty immediately.

(*b*) *Thermal burns*. When an individual suffers burns of the face from a fire, the eyes will close quickly due to extreme heat. This reaction is a natural reflex to protect the eyeballs; however, the eyelids remain exposed and are frequently burned. If a casualty receives burns of the eyelids or face—

- **DO NOT** apply a dressing.
- **DO NOT** touch.
- **SEEK** medical assistance immediately.

(*c*) *Light burns*. Exposure to intense light can burn an individual. Infrared rays, eclipse light (if the casualty has looked directly at the sun), or laser burns cause injuries of the exposed eyeball. Ultraviolet rays from arc welding can cause a superficial burn to the surface of the eye. These injuries are generally not painful but may cause permanent damage to the eyes. Immediate first aid is usually not required. Loosely bandaging the eyes may make the casualty more comfortable and protect his eyes from further injury caused by exposure to other bright lights or sunlight.

CAUTION

With impaled objects or significant sized foreign bodies, both eyes are usually bandaged to help secure the foreign body in the injured eye. In a battlefield environment, leave the uninjured eye uncovered so that the casualty can see.

c. *Side-of-Head or Cheek Wound*. Facial injuries to the side of the head or the cheek may bleed profusely (Figure 3-26). Prompt action is necessary to ensure that the airway remains open and also to control the bleeding. It may be necessary to apply a dressing. To apply a dressing—

(1) Remove the dressing from its wrapper.

(2) Grasp the tails in both hands.

(3) Hold the dressing directly over the wound with the white side down and place it directly on the wound (Figure 3-27A). **DO NOT** touch the white (sterile) side of the dressing or allow anything except the wound to come in contact with it.

(4) Hold the dressing in place with one hand (the casualty may assist if able). Wrap the top tail over the top of the head and bring it down in front of the ear (on the side opposite the wound), under the chin (Figure 3-27B) and up over the dressing to a point just above the ear (on the wound side).

Figure 3-26. Side of head or cheek wound.

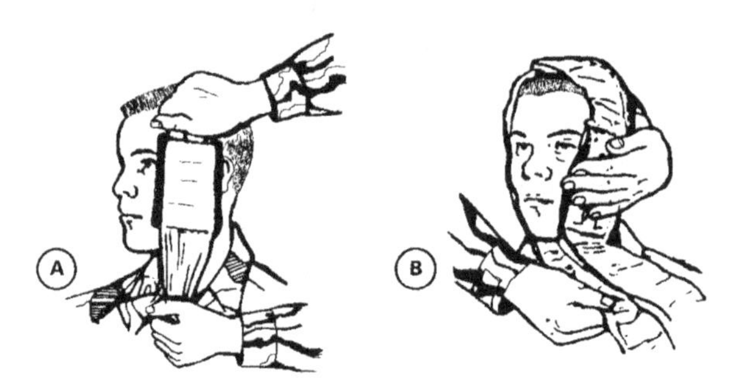

Figure 3-27. Dressing placed directly on wound. Top tail wrapped over top of head, down in front of ear, and under chin (Illustrated A—B).

NOTE

When possible, avoid covering the casualty's ear with the dressing, as this will decrease his ability to hear.

(5) Bring the second tail under the chin, up in front of the ear (on the side opposite the wound), and over the head to meet the other tail (on the wounded side) (Figure 3-28).

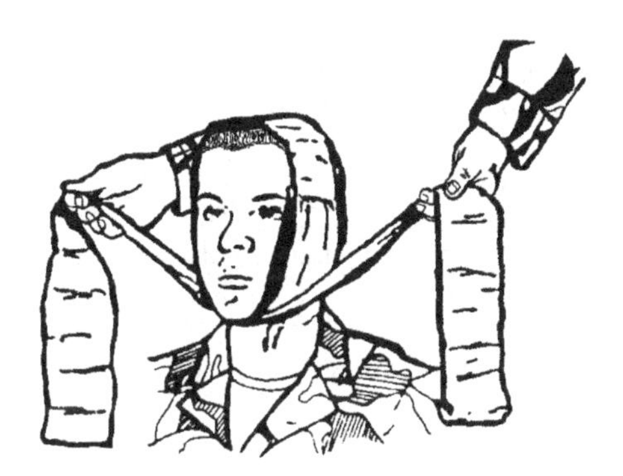

Figure 3-28. Bringing second tail under the chin.

(6) Cross thc two tails (on the wound side) (Figure 3-29) and bring one end across the forehead (above the eyebrows) to a point just in front of the opposite ear (on the uninjured side).

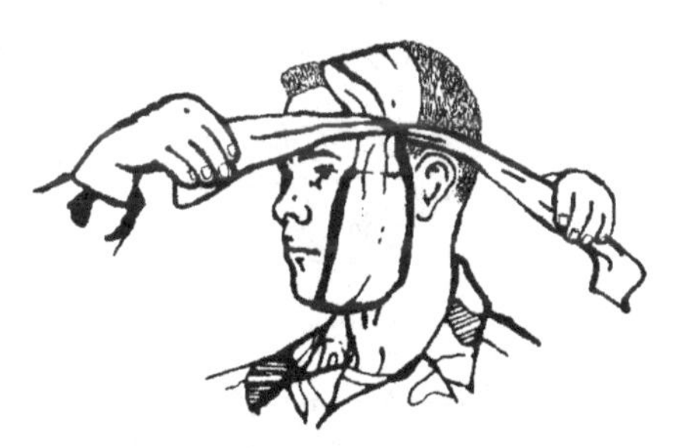

Figure 3-29. Crossing the tails on the side of the wound.

(7) Wrap the other tail around the back of the head (at the base of the skull), and tie the two ends just in front of the ear on the uninjured side with a square knot (Figure 3-30).

Figure 3-30. Tying the tails of the dressing in a square knot.

d. Ear Injuries. Lacerated (cut) or avulsed (torn) ear tissue may not, in itself, be a serious injury. Bleeding, or the drainage of fluids from the ear canal, however, may be a sign of a head injury, such as a skull fracture. **DO NOT** attempt to stop the flow from the inner ear canal nor put anything into the ear canal to block it. Instead, you should cover the ear lightly with a dressing. For minor cuts or wounds to the external ear, apply a cravat bandage as follows:

(1) Place the middle of the bandage over the ear (Figure 3-31A).

(2) Cross the ends, wrap them in opposite directions around the head, and tie them (Figures 3-31B and 3-31C).

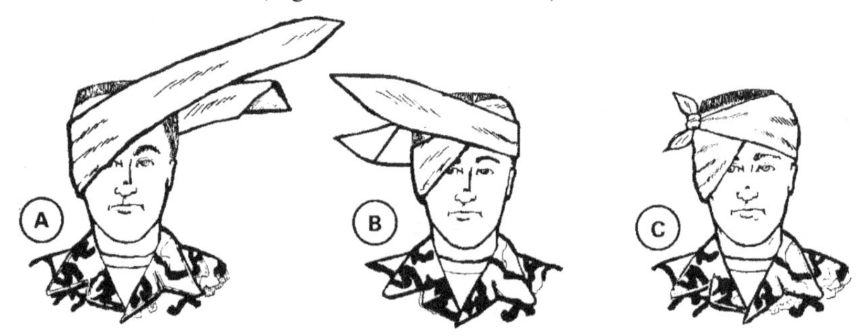

Figure 3-31. Applying cravat bandage to ear (Illustrated A—C).

(3) If possible, place some dressing material between the back of the ear and the side of the head to avoid crushing the ear against the head with the bandage.

e. Nose Injuries. Nose injuries generally produce bleeding. The bleeding may be controlled by placing an ice pack (if available) over the nose, or pinching the nostrils together. The bleeding may also be controlled by placing torn gauze (rolled) between the upper teeth and the lip.

CAUTION

DO NOT attempt to remove objects inhaled into the nose. An untrained person who removes such an object could worsen the casualty's condition and cause permanent injury.

f. Jaw Injuries. Before applying a bandage to a casualty's jaw, remove all loose or free-floating foreign material from the casualty's mouth.

If the casualty is unconscious, check for obstructions in the airway and remove if possible. If there is profuse bleeding in the oral cavity, the cavity may require loose packing with soft bandaging material (for example: Kerlix™ gauze) prior to applying a bandage. Care should be taken to avoid occluding the airway. When applying the bandage, allow the jaw enough freedom to permit passage of air and drainage from the mouth.

(1) *Apply bandages attached to field first aid dressing to the jaw.* After dressing the wound, apply the bandages using the same technique illustrated in Figure 3-32A—C.

NOTE

The dressing and bandaging procedure outlined for the jaw serves a twofold purpose. In addition to stopping the bleeding and protecting the wound, it also immobilizes a fractured jaw.

(2) *Apply a cravat bandage to the jaw.*

(*a*) Place the bandage under the chin and pull its ends upward. Adjust the bandage to make one end longer than the other (Figure 3-32A).

(*b*) Take the longer end over the top of the head to meet the short end at the temple and cross the ends over (Figure 3-32B).

(*c*) Take the ends in opposite directions to the other side of the head and tie them over the part of the bandage that was applied first (Figure 3-32C).

Figure 3-32. Applying a cravat bandage to jaw (Illustrated A—C).

NOTE

The cravat bandage technique is used to immobilize a fractured jaw or to maintain a sterile dressing that does not have tail bandages attached.

3-11. Shoulder Bandage

a. To apply bandages attached to the field first aid dressing—

(1) Take one bandage across the chest and the other across the back and under the arm opposite the injured shoulder.

(2) Tie the ends with a square knot (Figure 3-33).

Figure 3-33. Shoulder bandage.

b. To apply a cravat bandage to the shoulder or armpit—

(1) Make an extended cravat bandage by using two triangular bandages (Figure 3-34A); place the end of the first triangular bandage along the base of the second one (Figure 3-34B).

(2) Fold the two bandages into a single extended bandage (Figure 3-34C).

(3) Fold the extended bandage into a single cravat bandage (Figure 3-34D). After folding, secure the thicker part (overlap) with two or more safety pins (Figure 3-34E).

(4) Place the middle of the cravat bandage under the armpit so that the front end is longer than the back end and safety pins are on the outside (Figure 3-34F).

(5) Cross the ends on top of the shoulder (Figure 3-34G).

(6) Take one of the bandage ends across the back and under the arm on the opposite side and the other end across the chest. Tie the ends (Figure 3-34H).

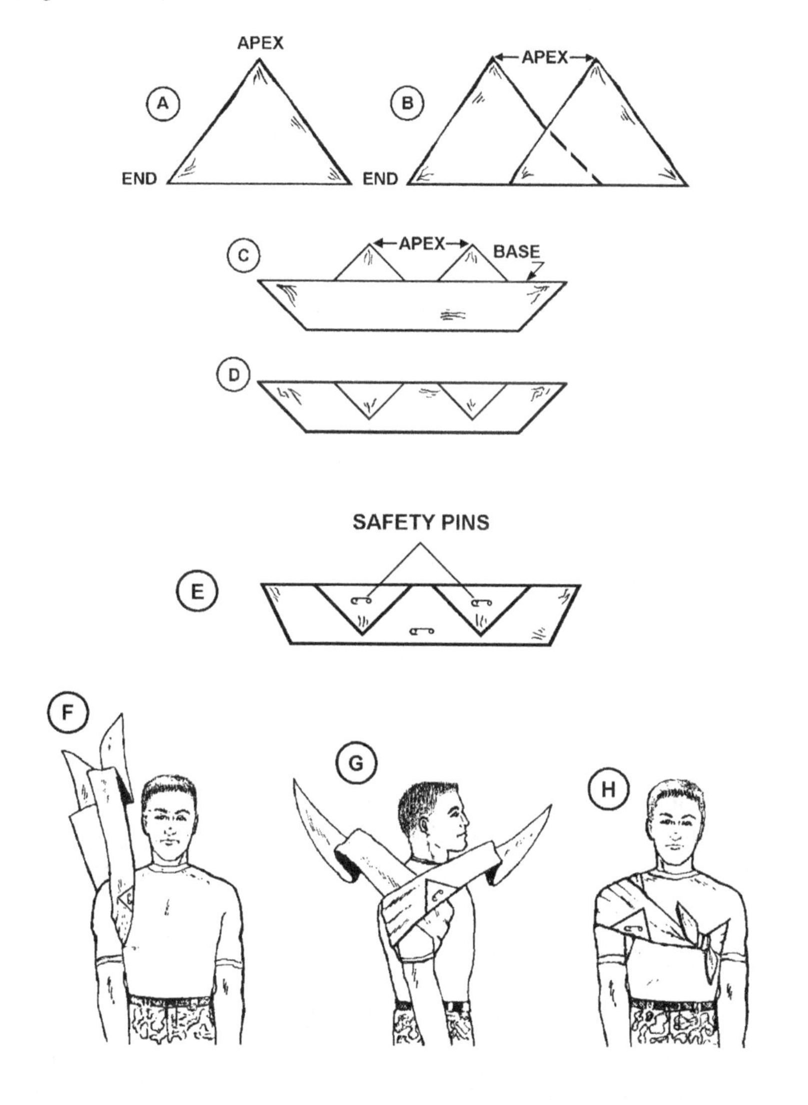

Figure 3-34. Extended cravat bandage applied to shoulder or armpit (Illustrated A—H).

Be sure to place sufficient wadding in the armpit. **DO NOT** tie the cravat bandage too tightly. Avoid compressing the major blood vessels in the armpit.

3-12. Elbow Bandage

To apply a cravat bandage to the elbow—

a. Bend the arm at the elbow and place the middle of the cravat at the point of the elbow bringing the ends upward (Figure 3-35A).

b. Bring the ends across, extending both downward (Figure 3-35B).

c. Take both ends around the arm and tie them with a square knot at the front of the elbow (Figure 3-35C).

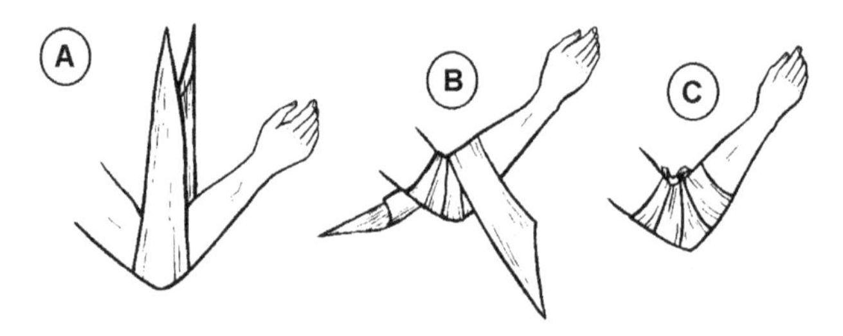

Figure 3-35. Elbow bandage (Illustrated A—C).

CAUTION

If an elbow fracture is suspected, **DO NOT** bend the elbow; bandage it in the position found.

3-13. Hand Bandage

a. To apply a triangular bandage to the hand—

(1) Place the hand in the middle of the triangular bandage with the wrist at the base of the bandage (Figure 3-36A). Ensure that the fingers are separated with absorbent material to prevent chafing and irritation of the skin.

(2) Place the apex over the fingers and tuck any excess material into the pleats on each side of the hand (Figure 3-36B).

(3) Cross the ends on top of the hand, take them around the wrist, and tie them (Figures 3-36C—E) with a square knot.

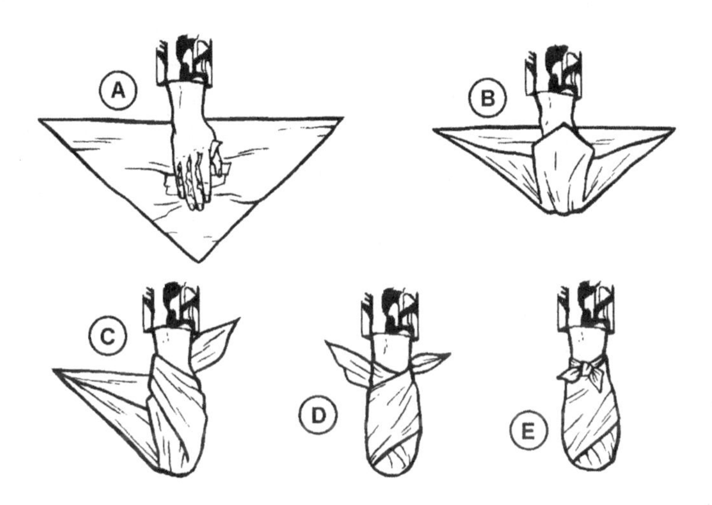

Figure 3-36. Triangular bandage applied to hand (Illustrated A—E).

b. To apply a cravat bandage to the palm of the hand—

(1) Lay the middle of the cravat over the palm of the hand with the ends hanging down on each side (Figure 3-37A).

(2) Take the end of the cravat at the little finger across the back of the hand, extending it upward over the base of the thumb; then bring it downward across the palm (Figure 3-37B).

(3) Take the thumb end across the back of the hand, over the palm, and through the hollow between the thumb and palm (Figure 3-37C).

(4) Take the ends to the back of the hand and cross them; then bring them up over the wrist and cross them again (Figure 3-37D).

(5) Bring both ends down and tie them with a square knot on top of the wrist (Figure 3-37E—F).

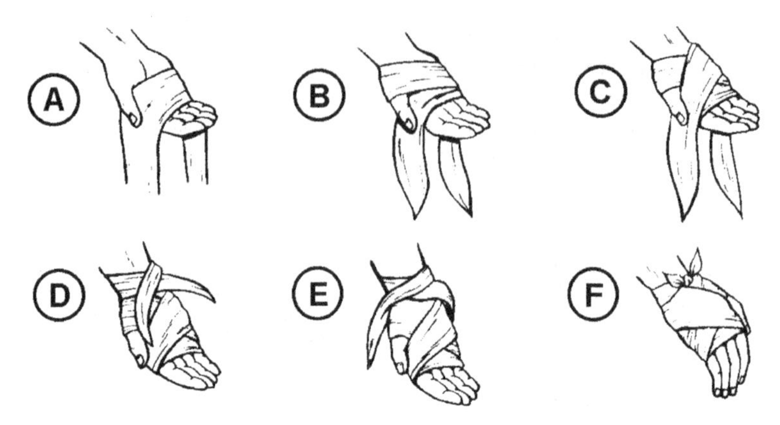

Figure 3-37. Cravat bandage applied to palm of hand (Illustrated A—F).

3-14. Leg (Upper and Lower) Bandage

To apply a cravat bandage to the leg—

a. Place the center of the cravat over the dressing (Figure 3-38A).

b. Take one end around and up the leg in a spiral motion and the other end around and down the leg in a spiral motion, overlapping part of each preceding turn (Figure 3-38B).

c. Bring both ends together and tie them (Figure 3-38C) with a square knot.

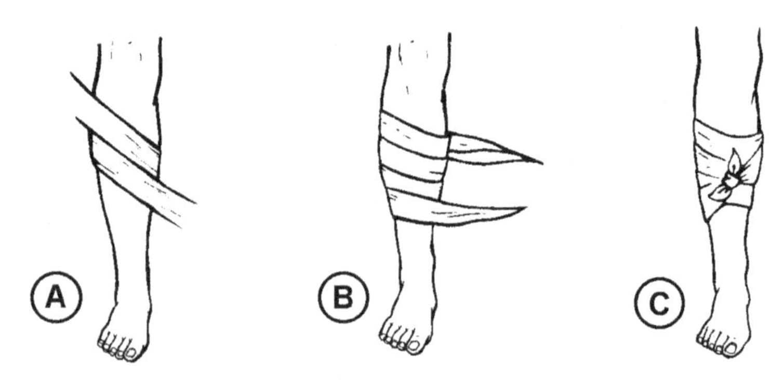

Figure 3-38. Cravat bandage applied to leg (Illustrated A—C).

3-15. Knee Bandage

To apply a cravat bandage to the knee as illustrated in Figure 3-39, use the same technique applied in bandaging the elbow.

> **CAUTION**
>
> If a fracture of the kneecap is suspected, **DO NOT** bend the knee; bandage it in the position found.

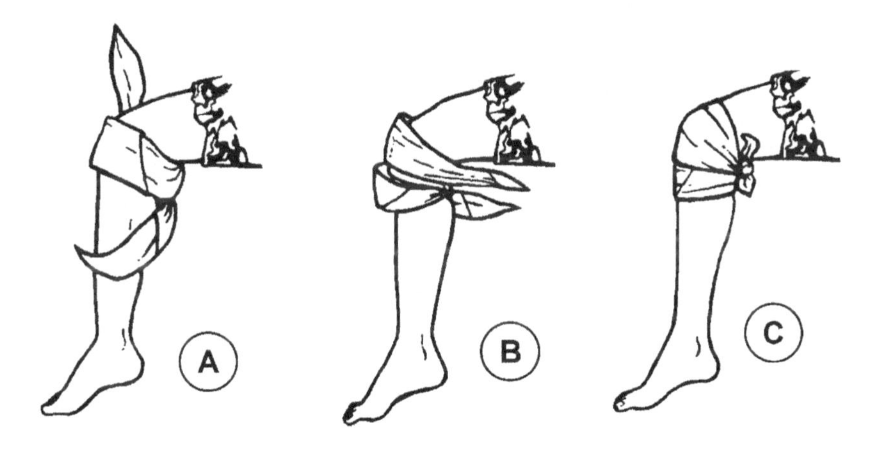

Figure 3-39. Cravat bandage applied to knee (Illustrated A—C).

3-16. Foot Bandage

To apply a triangular bandage to the foot—

a. Place the foot in the middle of the triangular bandage with the heel well forward of the base (Figure 3-40A). Ensure that the toes are separated by absorbent material to prevent chafing and irritation of the skin.

b. Place the apex over the top of the foot and tuck any excess material into the pleats on each side of the foot (Figure 3-40B).

c. Cross the ends on top of the foot, take them around the ankle, and tie them at the front of the ankle (Figure 3-40C—E).

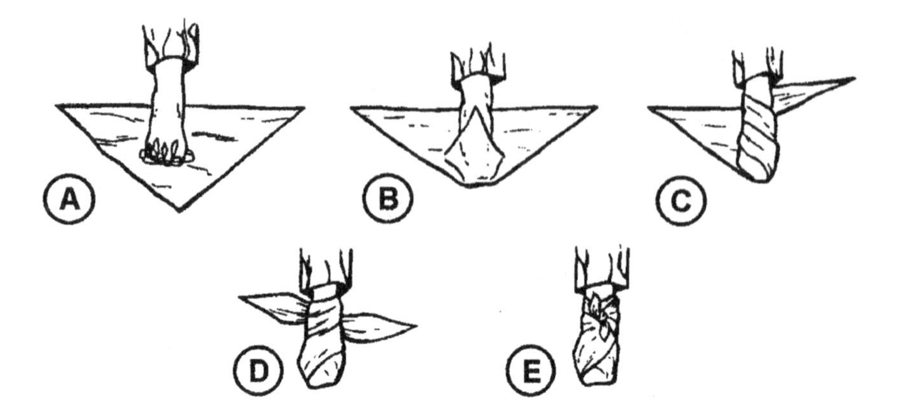

Figure 3-40. Triangular bandage applied to foot (Illustrated A—E).

CHAPTER 4

FIRST AID FOR FRACTURES

4-1. General

A fracture is any break in the continuity of a bone. Fractures can cause total disability or in some cases death by severing vital organs and/or arteries. On the other hand, they can most often be treated so there is a complete recovery. The potential for recovery depends greatly upon the first aid the individual receives before he is moved. First aid includes immobilizing the fractured part in addition to applying lifesaving measures when necessary. The basic splinting principle is to immobilize the joints above and below the fracture.

4-2. Kinds of Fractures

Figure 4-1 depicts types of fractures.

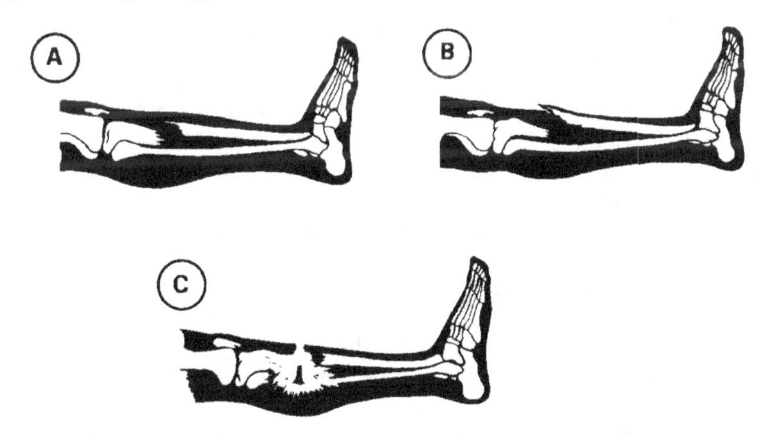

Figure 4-1. Types of fractures (Illustrated A—C).

a. *Closed Fracture* (Figure 4-1A). A closed fracture is a broken bone that does not break the overlying skin. The tissue beneath the skin may be damaged. A *dislocation* is when a joint, such as a knee, ankle, or shoulder, is not in the proper position. A *sprain* is when the connecting tissues of the joints have been torn. Dislocations and sprains (swelling, possible deformity, and discoloration) should be treated as closed fractures.

b. *Open Fracture* (Figure 4-1B and 4-1C). An open fracture is a broken bone that breaks (pierces) the overlying skin. The broken bone may come through the skin or a missile such as a bullet or shell fragment may go through the flesh and break the bone.

NOTE

An open fracture is contaminated and subject to infection.

4-3. Signs and Symptoms of Fractures

Indications of a fracture are deformity, tenderness, swelling, pain, inability to move the injured part, protruding bone, bleeding, or discolored skin at the injury site. A sharp pain when the service member attempts to move the part is also a sign of a fracture.

WARNING

DO NOT encourage the casualty to move the injured part in order to identify a fracture since such movement could cause further damage to surrounding tissues and promote shock. If you are not sure whether a bone is fractured, care for the injury as a fracture. At the site of the fracture, the bone ends are sharp and could cause vessel (artery and/or vein) damage.

4-4. Purposes of Immobilizing Fractures

A fracture is immobilized to prevent the sharp edges of the bone from moving and cutting tissue, muscle, blood vessels, and nerves. This reduces pain and helps prevent or control shock. In a closed fracture, immobilization keeps bone fragments from causing an open wound, which can become contaminated and subject to infection.

4-5. Splints, Padding, Bandages, Slings, and Swathes

a. Splints. Splints may be improvised from such items as boards, poles, sticks, tree limbs, or cardboard. If nothing is available for a splint, the chest wall can be used to immobilize a fractured arm and the uninjured leg can be used to immobilize (to some extent) the fractured leg.

b. Padding. Padding may be improvised from such items as a jacket, blanket, poncho, shelter half, or leafy vegetation.

c. Bandages. Bandages may be improvised from belts, rifle slings, kerchiefs, or strips torn from clothing or blankets. Narrow materials such as wire or cord should not be used to secure a splint in place. The application of wire and/or narrow material to an extremity could cause tissue damage and a tourniquet effect.

d. Slings. A sling is a bandage suspended from the neck to support an upper extremity. If a bandage is not available, a sling can be improvised by using the tail of a coat or shirt or pieces of cloth torn from such items as clothing and blankets. The triangular bandage is ideal for this purpose. Remember that the casualty's hand should be higher than his elbow, and the fingers should be showing at all times. The sling should be applied so that the supporting pressure is on the uninjured side.

e. Swathes. Swathes are any bands (pieces of cloth or load bearing equipment [LBE]) that are used to further immobilize a splinted fracture. Triangular and cravat bandages are often used and are called *swathe bandages*. The purpose of the swathe is to immobilize; therefore, the swathe bandage is placed above and/or below the fracture—not over it.

4-6. Procedures for Splinting Suspected Fractures

Before beginning first aid procedures for a fracture, gather whatever splinting materials are available. Ensure that splints are long enough to immobilize the joint above and below the suspected fracture. If possible, use at least four ties (two above and two below the fracture) to secure the splints. The ties should be square knots and should be tied away from the body on the splint. Distal pulses of the affected extremity should be checked before and after the application of the splint.

a. Evaluate the Casualty. Be prepared to perform any necessary lifesaving measures. Monitor the casualty for development of conditions that may require you to perform necessary lifesaving measures.

WARNING

Unless there is immediate life-threatening danger, such as a fire or an explosion, DO NOT move the casualty with a suspected back or neck injury. Improper movement may cause permanent paralysis or death.

WARNING

In a chemical environment, DO NOT remove any protective clothing. Apply the dressings and splints over the garments.

b. Locate the Site of the Suspected Fracture.

(1) Ask the casualty for the location of the injury.

- Does he have any pain?
- Where is it tender?
- Can he move the extremity?

NOTE

With the presence of an obvious deformity, do not make the casualty move extremity.

(2) Look for an unnatural position of the extremity.

(3) Look for a bone sticking out (protruding).

c. Prepare the Casualty for Splinting the Suspected Fracture.

(1) Reassure the casualty. Tell him that you will be providing first aid for him and that medical help is on the way.

(2) Loosen any tight or binding clothing.

(3) Remove all jewelry from the injured part and place it in the casualty's pocket. Tell the casualty you are doing this because if the jewelry is not removed and swelling occurs later, he may not be able to get it off and further bodily injury could result.

(4) Boots should not be removed from the casualty unless they are needed to stabilize a neck injury or there is actual bleeding from the foot.

d. Gather Splinting Materials. If standard splinting materials (splints, padding, and cravats) are not available, gather improvised materials. If splinting material is not available and the suspected fracture **CANNOT** be splinted, then swathes, or a combination of swathes and slings can be used to immobilize the extremity.

e. *Pad the Splints*. Pad the splints where they touch any bony part of the body, such as the elbow, wrist, knee, ankle, crotch, or armpit areas. Padding prevents excessive pressure on the area, which could lead to circulation problems.

f. *Check the Circulation Below the Site of the Injury*.

(1) Note any pale, white, or bluish-gray color of the skin, which may indicate impaired circulation. Circulation can also be checked by depressing the toe or fingernail beds and observing how quickly the color returns. A slower return of color to the injured side when compared with the uninjured side indicates a problem with circulation. The fingernail bed is the method to use to check the circulation in a dark-skinned casualty.

(2) Check the temperature of the injured extremity. Use your hand to compare the temperature of the injured side with the uninjured side. The body area below the injury may be colder to the touch indicating poor circulation.

(3) Question the casualty about the presence of numbness, tightness, cold, or tingling sensations.

WARNING

Casualties with fractures of the extremities may show impaired circulation, such as numbness, tingling, cold or pale to bluish skin tone. These casualties should be evacuated by medical personnel and treated as soon as possible. Prompt medical treatment may prevent possible loss of the limb.

WARNING

If it is an open fracture and the bone is protruding from the skin, *DO NOT ATTEMPT TO PUSH THE BONE BACK UNDER THE SKIN.* Apply a field dressing over the wound to protect the area.

g. Apply the Splint in Place.

(1) Splint the fracture in the position found. **DO NOT** attempt to reposition or straighten the injury. If it is an open fracture, stop the bleeding and protect the wound. Cover all wounds with field dressings before applying a splint. Remember to use the casualty's field dressing, not your own.

(2) Place one splint on each side of the fracture. Make sure that the splints reach, if possible, beyond the joints above and below the fracture.

(3) Tie the splints. Secure each splint in place above and below the fracture site with improvised (or actual) cravats. Improvised cravats, such as strips of cloth, belts, or whatever else you have, may be used. With minimal motion to the injured areas, place and tie the splints with the bandages. Push cravats through and under the natural body curvatures, and then gently position improvised cravats and tie in place. Use square knots. Tie all knots on the splint away from the casualty (Figure 4-2). **DO NOT** tie cravats directly over the suspected fracture site.

Figure 4-2. Square knots tied away from casualty.

h. Check the Splint for Tightness.

(1) **CHECK** to be sure that bandages are tight enough to securely hold splinting materials in place, but not so tight that circulation is impaired.

(2) **RECHECK** the circulation after application of the splint. Check the skin color and temperature. This is to ensure that the bandages holding the splint in place have not been tied too tightly. A fingertip check can be made by inserting the tip of the finger between the bandaged knot and the skin.

(3) **MAKE** any necessary adjustment without allowing the splint to become ineffective.

i. Apply a Sling. An improvised sling may be made from any available nonstretching piece of cloth, such as a battle dress uniform (BDU) shirt or trousers, poncho, or shelter half. Slings may also be improvised using the tail of a coat, belt, or a piece of cloth. Figure 4-3 depicts a shirttail used for support. A trousers belt or LBE may also be used for support (Figure 4-4). A sling should place the supporting pressure on the casualty's uninjured side. The supported arm should have the hand positioned slightly higher than the elbow showing the fingers.

Figure 4-3. Shirttail used for support.

Figure 4-4. Belt used for support.

(1) Insert the splinted arm in the center of the sling (Figure 4-5).

Figure 4-5. Arm inserted in center of improvised sling.

(2) Bring the ends of the sling up and tie them at the side (or hollow) of the neck on the uninjured side (Figure 4-6).

Figure 4-6. Ends of improvised sling tied to side of neck.

(3) Twist and tuck the corner of the sling at the elbow (Figure 4-7).

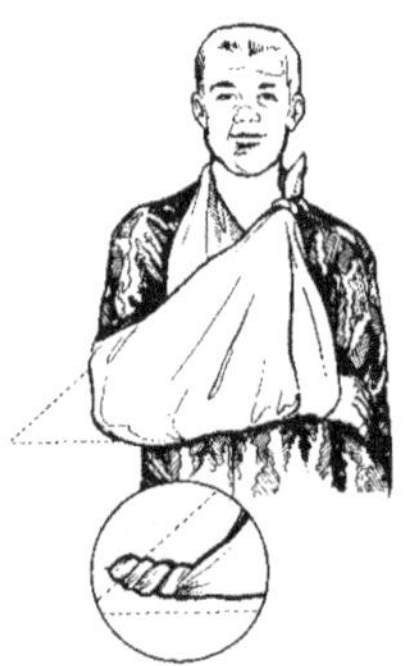

Figure 4-7. Corner of sling twisted and tucked at elbow.

j. Apply a Swathe. You may use any large piece of cloth, service member's belt, or pistol belt, to improvise a swathe.

WARNING

The swathe should not be placed directly on top of the injury, but positioned either above or below the fracture site.

(1) Apply swathes to the injured arm by wrapping the swathe over the injured arm, around the casualty's back, and under the arm on the uninjured side. Tie the ends on the uninjured side (Figure 4-8).

Figure 4-8. Arm immobilized with strip of clothing.

(2) A swathe is applied to an injured leg by wrapping the swathe around both legs and securing it on the uninjured side.

k. Seek Medical Assistance. Notify medical personnel, watch closely for development of life-threatening conditions and/or impaired circulation to the injured extremity. (Refer to Chapter 1 for additional information on life-threatening conditions.)

4-7. Upper Extremity Fractures

Figures 4-9 through 4-17 show how to apply slings, splints, and cravats (swathes) to immobilize and support fractures of the upper extremities. *Although the padding is not visible in some of the illustrations, it is always preferable to apply padding along the injured part for the length of the splint and especially where it touches any bony parts of the body.*

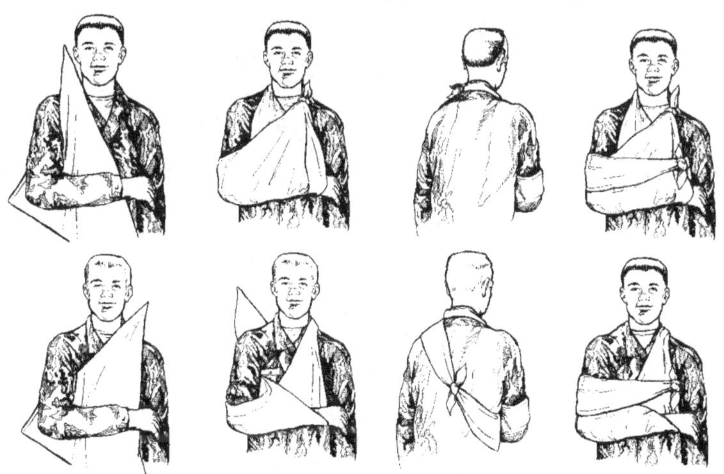

Figure 4-9. Application of triangular bandage to form sling (two methods).

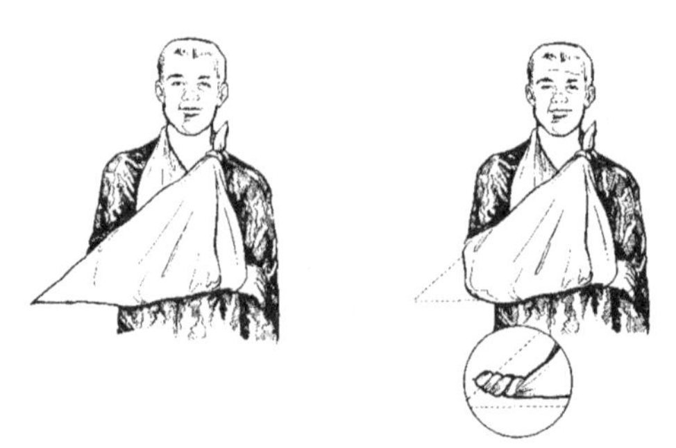

Figure 4-10. Completing sling sequence by twisting and tucking the corner of the sling at the elbow.

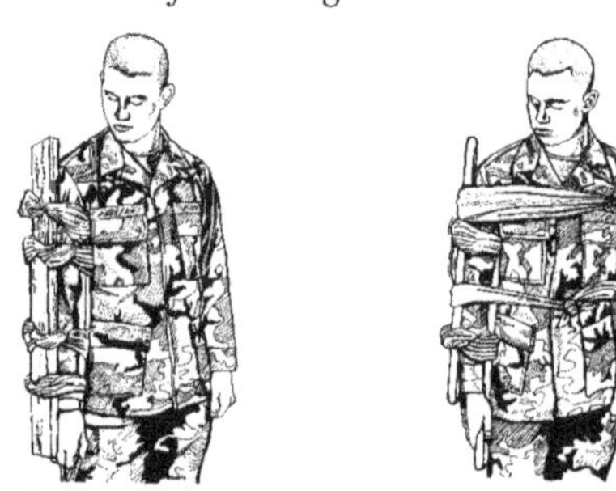

Figure 4-11. Board splints applied to fractured elbow when elbow is not bent (two methods).

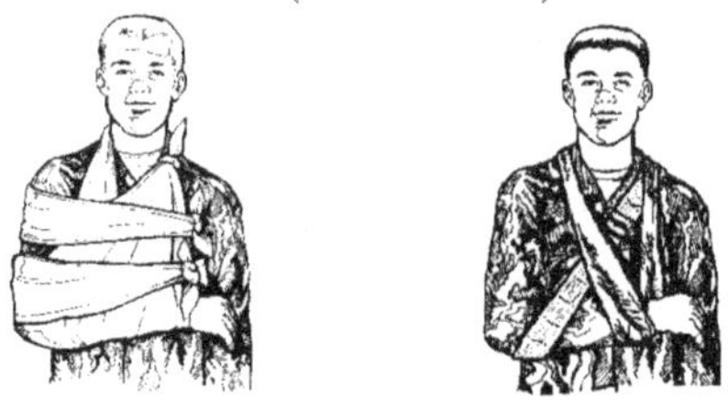

Figure 4-12. Chest wall used as splint for upper arm fracture when no splint is available.

Figure 4-13. Chest wall, sling, and cravat used to immobilize fractured elbow when elbow is bent.

Figure 4-14. Board splint applied to fractured forearm.

Figure 4-15. Fractured forearm or wrist splinted with sticks and supported with tail of shirt and strips of material.

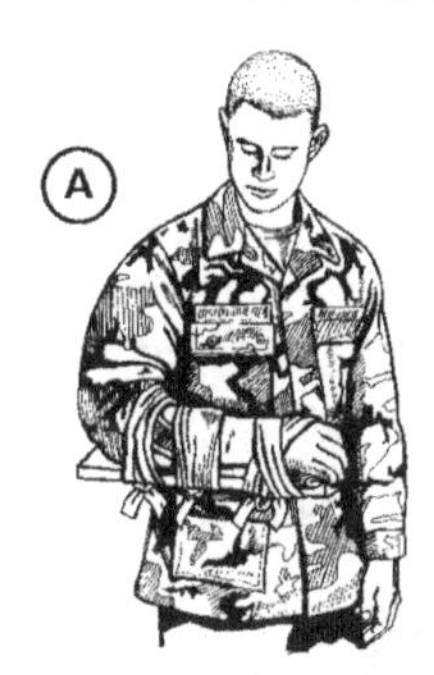

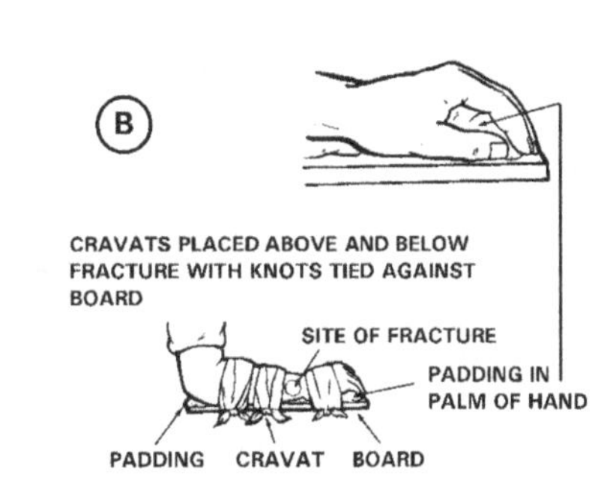

Figure 4-16. Board splint applied to fractured wrist and hand (Illustrated A—B).

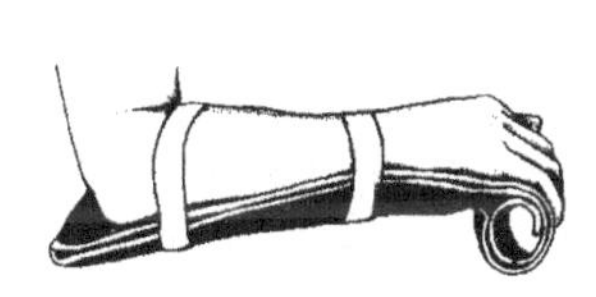
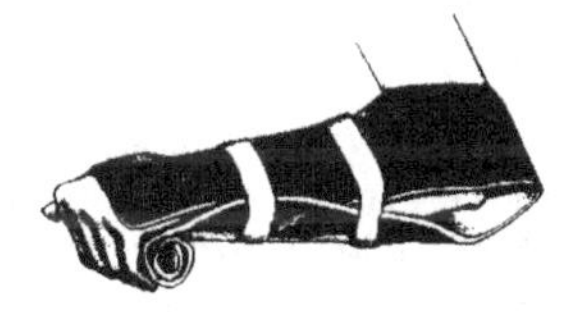

Figure 4-17. SAM® splint applied to fractured wrist or forearm.

4-8. Lower Extremity Fractures

Figures 4-18 through 4-24 show how to apply splints to immobilize fractures of the lower extremities. *Although padding is not visible in some of the illustrations, it is always preferable to apply padding along the injured part for the length of the splint and especially where it touches any bony parts of the body.*

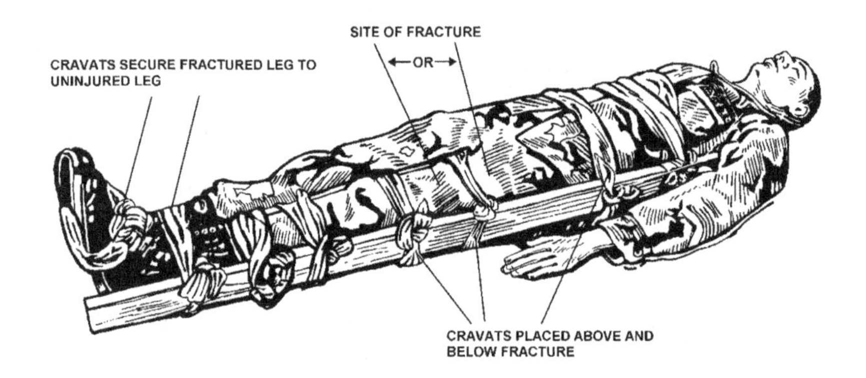

Figure 4-18. Board splints applied to fractured hip or thigh.

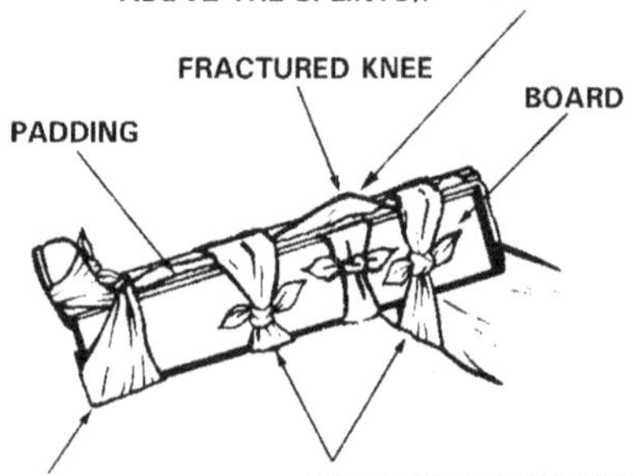

Figure 4-19. Board splint applied to fractured or dislocated knee.

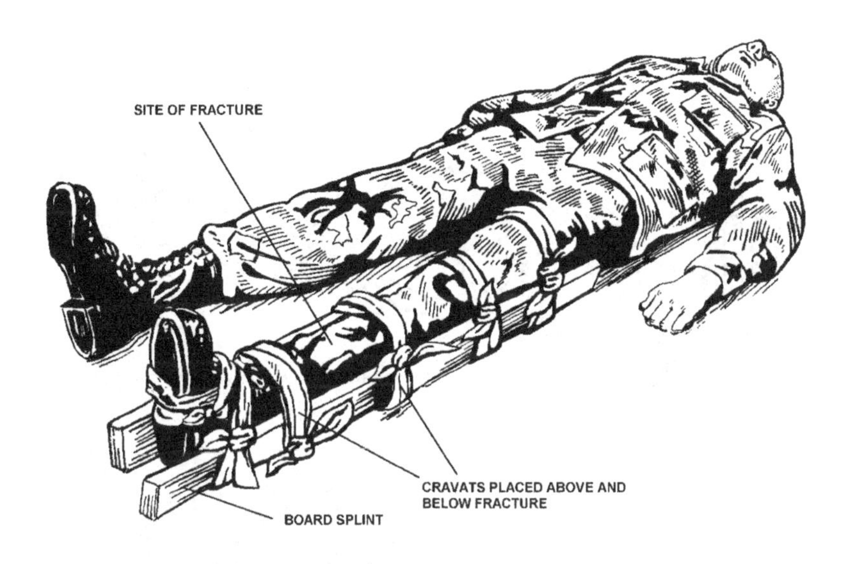

Figure 4-20. Board splints applied to fractured lower leg or ankle.

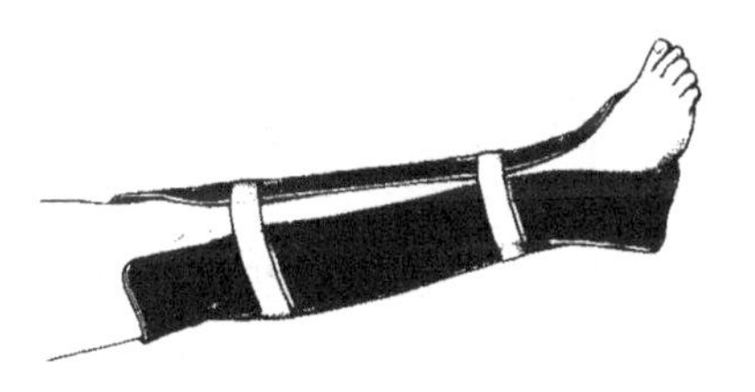

Figure 4-21. SAM® splint applied to fractured lower leg or ankle.

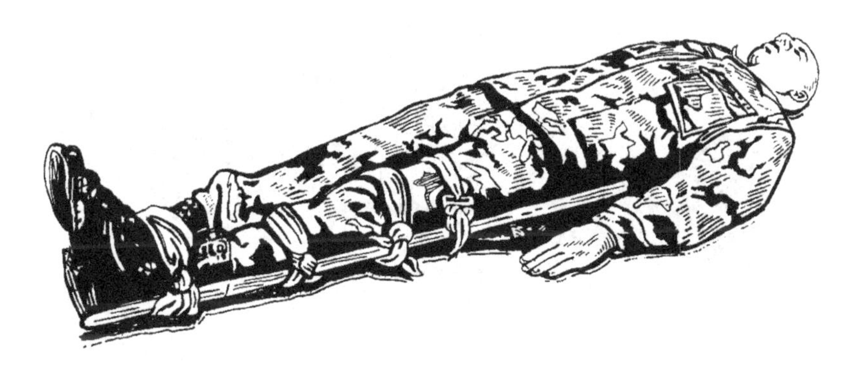

Figure 4-22. Improvised splints applied to fractured lower leg or ankle.

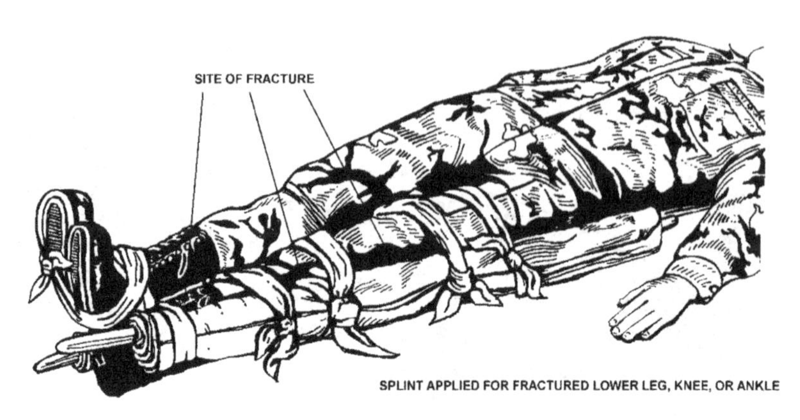

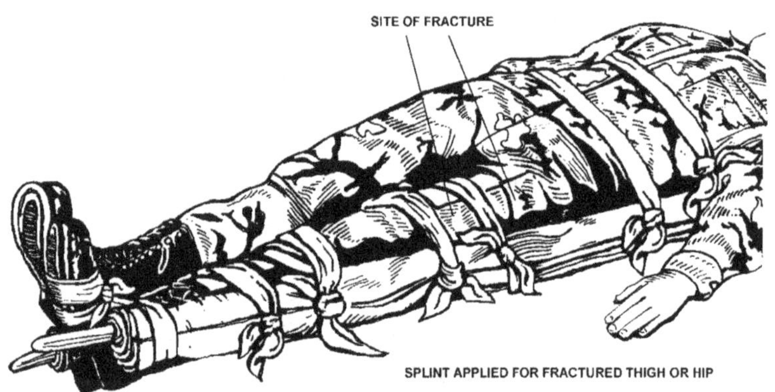

Figure 4-23. Poles rolled in a blanket and used as splints applied to fractured lower extremity.

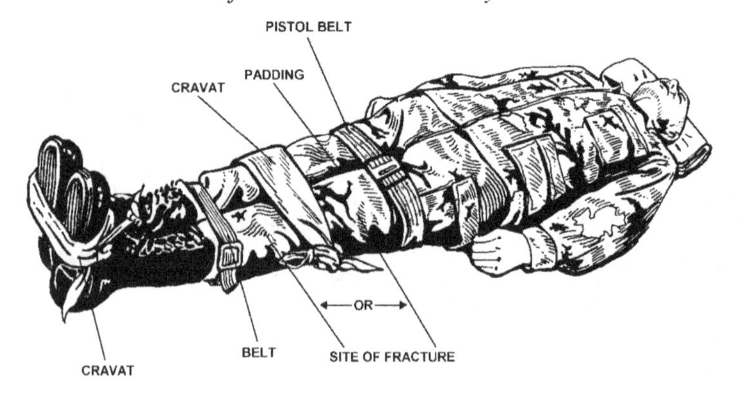

Figure 4-24. Uninjured leg used as splint for fractured leg (anatomical splint).

4-9. Jaw, Collarbone, and Shoulder Fractures

a. Apply a cravat to immobilize a fractured jaw as illustrated in Figure 4-25. Direct all bandaging support to the top of the casualty's head, not to the back of his neck. If incorrectly placed, the bandage will pull the casualty's jaw back and interfere with his breathing.

Figure 4-25. Fractured jaw immobilized.

WARNING

Casualties with lower jaw (mandible) fractures cannot be laid flat on their backs because facial muscles will relax and may cause an airway obstruction.

b. Apply two belts, a sling, and a cravat to immobilize a fractured collarbone, as illustrated in Figure 4-26.

Figure 4-26. Application of belts, sling, and cravat to immobilize a fractured collarbone.

c. Apply a sling and a cravat to immobilize a fractured or dislocated shoulder, using the technique illustrated in Figure 4-27.

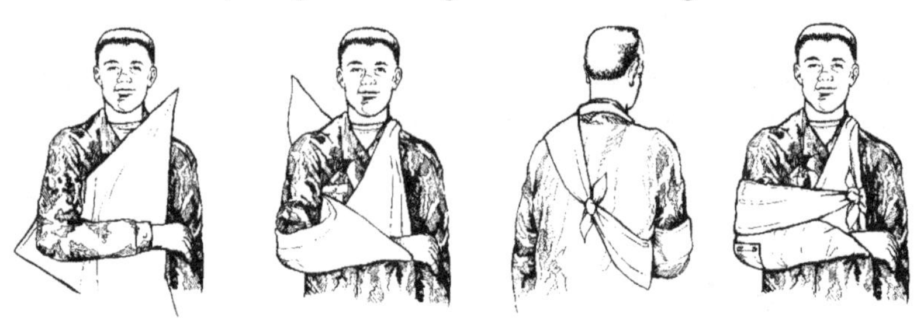

Figure 4-27. Application of sling and cravat to immobilize a fractured or dislocated shoulder.

4-10. Spinal Column Fractures

It is often impossible to be sure a casualty has a fractured spinal column. Be suspicious of any back injury, especially if the casualty has fallen or if his back has been sharply struck or bent. If a casualty has received such an injury and does not have feeling in his legs or cannot move them, you can be reasonably sure that he has a severe back injury, which should be managed as a fracture. Remember, that the possibility of a neck fracture or injury to the back should always be suspected, and it is often impossible to be sure if a casualty has a fractured spinal column. If the spine is fractured, bending it can cause the sharp bone fragments to bruise or cut the spinal cord and result in permanent paralysis or death (Figure 4-28A). The spinal column must maintain normal spinal position at the lower back (lumbar region) to help remove pressure from the spinal cord.

a. If the casualty is not to be transported until medical personnel arrive—

- Caution him not to move. Ask him if he is in pain or if he is unable to move any part of his body.

- Leave him in the position in which he is found. **DO NOT** move any part of his body, unless he is in imminent danger.

- If the casualty is lying face up, slip a blanket or other supporting material under the arch of his lower back to help support the spine in a normal position (Figure 4-28B). Take care not to place so much bulky padding as to cause potential damage by causing undo pressure on the spine. If he is lying face down, **DO NOT** put anything under any part of his body.

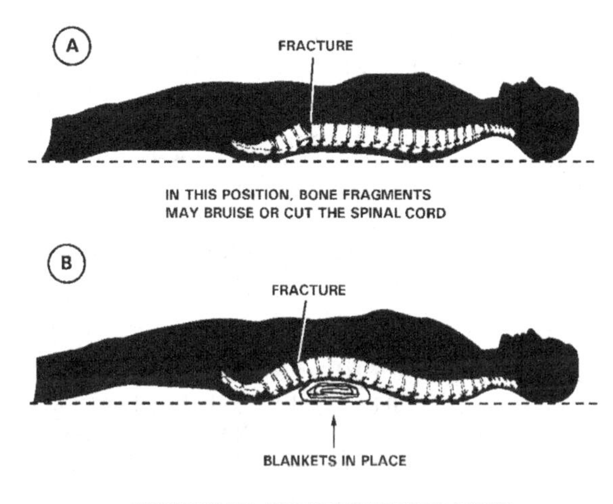

Figure 4-28. Spinal column must maintain a normal spine position.

b. If the casualty must be transported to a safe location before medical personnel arrive and if the casualty is in a—

- Face-up position, transport him by litter or use a firm substitute, such as a wide board or a door longer than his height. Loosely tie the casualty's wrists together over his waistline, using a cravat or a strip of cloth. Tie his feet together to prevent the accidental dropping or shifting of his legs. Lay a folded blanket across the litter where the arch of his back is to be placed. Using a four-man team (Figure 4-29), place the casualty on the litter without bending his spinal column or his neck.

Figure 4-29. Placing face-up casualty with fractured back onto litter.

- The number two man positions himself at the casualty's head. His responsibility is to provide manual in-line (neutral) stabilization of the head and neck. The number three, and four men position themselves on one side of the casualty; all kneel on one knee along the side of the casualty. The number one man positions himself to the opposite side of the casualty (or can be on the same side of number three and four). The numbers two, three, and four men gently place their hands under the casualty. The number one man on the opposite side places his hands under the injured part to assist.

- When all four men are in position to lift, the number two man commands, "**PREPARE TO LIFT**" and then, "**LIFT**." All men, in unison, gently lift the casualty about 8 inches. Once the casualty is lifted, the number one man recovers and slides the litter under the casualty, ensuring that the blanket is in proper position. The number one man then returns to his original lift position (Figure 4-29).

- When the number two man commands, "**LOWER CASUALTY**," all men, in unison, gently lower the casualty onto the litter.

- Facedown position, he must be transported in this same position. The four-man team lifts him onto a regular or improvised litter, keeping the spinal column in a normal spinal position. If a regular litter is used, first place a folded blanket on the litter at the point where the chest will be placed.

4-11. Neck Fractures

A fractured neck is extremely dangerous. Bone fragments may bruise or cut the spinal cord just as they might in a fractured back.

a. If the casualty is not to be transported until medical personnel arrive—

- Caution him not to move. Moving may cause permanent injury or death.

- Leave the casualty in the position in which he is found. If his neck and head (cervical spine) are in an abnormal position, immediately immobilize the neck and head.

- Keep his head still, if the casualty is lying face up, raise his shoulders slightly, and slip a roll of cloth that has the bulk of a bath towel under his neck (Figure 4-31). The roll should be thick enough to arch his neck only slightly, leaving the back of his head on the ground. **DO NOT** bend his neck or head forward. **DO NOT** raise or twist his head. Immobilize the casualty's head (Figure 4-32). Do this by padding heavy objects (such as rocks or his boots filled with dirt, sand, gravel, or rock) and placing them on each side of his head. If it is necessary to use boots, after filling them, tie the top tightly or stuff with pieces of cloth to secure the contents.)

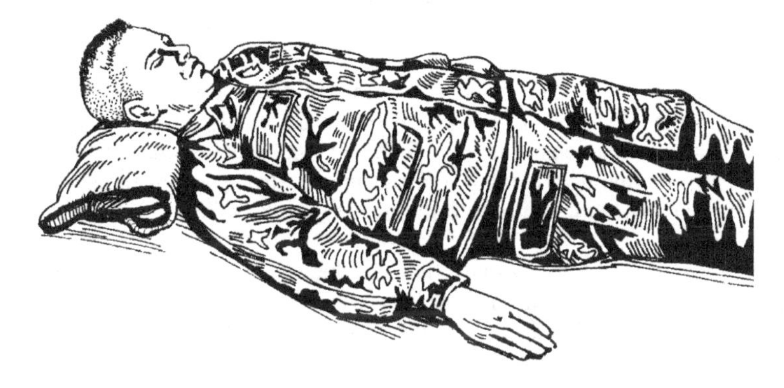

Figure 4-30. Casualty with roll of cloth (bulk) under neck.

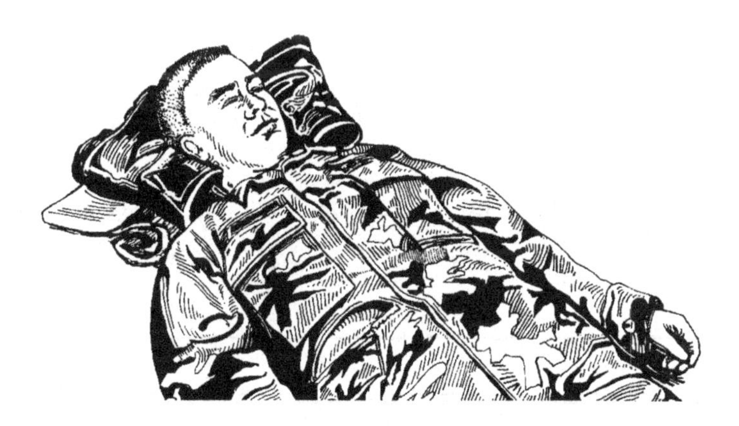

Figure 4-31. Immobilization of fractured neck.

- **DO NOT** move him if the casualty is lying face down. Immobilize the head and neck by padding heavy objects and placing them on each side of his head. DO NOT put a roll of cloth under the neck. **DO NOT** bend the neck or head, nor roll the casualty onto his back.

b. If the casualty must be prepared for transportation before medical personnel arrive—

- If the casualty has a fractured neck, at least two persons are needed because the casualty's head and trunk must be moved in unison. The two persons must work in close coordination (Figure 4-32) to avoid bending of the neck.

- A wide board is placed lengthwise beside the casualty. It should extend at least 4 inches beyond the casualty's head and feet (Figure 4-32A).

- If the casualty is lying face up, the number one man steadies the casualty's head and neck between his hands. At the same time, the number two man positions one foot and one knee against the board to prevent it from slipping. He then grasps the casualty underneath his shoulder and hip and gently slides him onto the board (Figure 4-32B).

- If the casualty is lying face down, the number one man steadies the casualty's head and neck between his hands, while the number two man gently rolls the casualty over onto the board (Figure 4-32C).

- The number one man continues to steady the casualty's head and neck. The number two man simultaneously raises the casualty's shoulders slightly, places padding under his neck, and immobilizes the casualty's head (Figures 4-32D—E).

- Any improvised supports are secured in position with a cravat or strip of cloth extended across the casualty's forehead and under the board (Figure 4-32D).

- The board is lifted onto a litter or blanket in order to transport the casualty (Figure 4-32E).

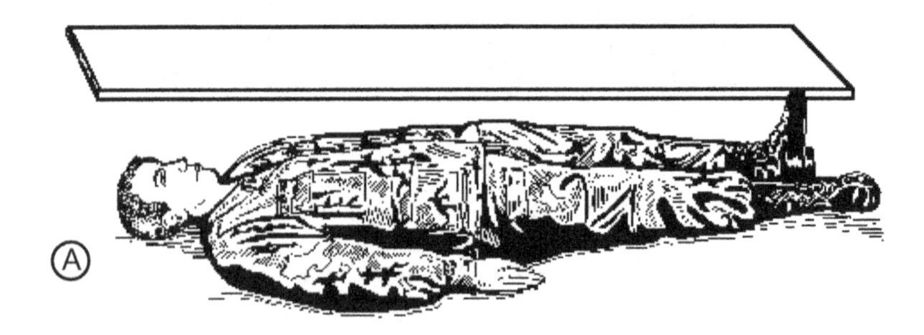

Figure 4-32. Preparing casualty with fractured neck for transportation (Illustrated A—E).

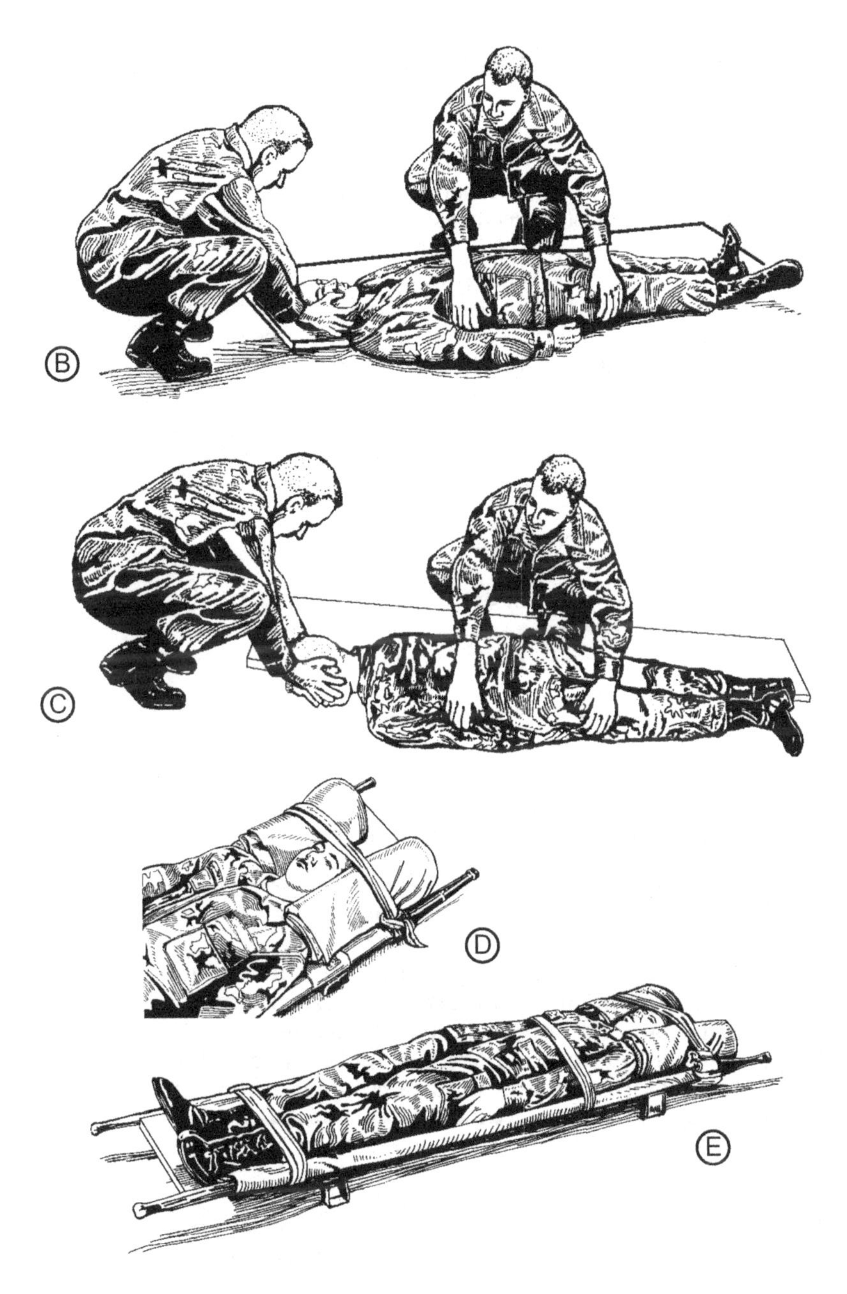

Figure 4-32. Preparing casualty with fractured neck for transportation (Illustrated A—E) (Continued).

CHAPTER 5

FIRST AID FOR CLIMATIC INJURIES

5-1. General

a. It is desirable, but not always possible, for an individual's body to become adjusted (acclimated) to an environment.

(1) The service members physical condition determines the amount of time their bodies need to adjust to the environment. Even those individuals in good physical condition need time before working or training in extremes of hot or cold weather. Climate-related injuries are usually preventable; prevention is both an individual and leadership responsibility.

(2) Several factors contribute to health and well-being in any environment—

- Diet.
- Sleep and rest.
- Exercise.
- Suitable clothing.

(3) Diet should be suited to an individual's needs in a particular climate. A special diet started for any purpose (such as weight reduction) should be done with appropriate medical supervision.

WARNING

Service members should use extreme caution when starting fad diets or taking over-the-counter herbal supplements. Medical records revealed that deaths and severe injuries occurred in individuals using dietary/ herbal supplements without medical monitoring.

NOTE

Weight loss and the use of weight loss supplements should be supervised by a trained health care provider.

(4) Specialized clothing and equipment (such as cold weather gear) for a specific environment should be obtained and used properly.

b. For information on the prevention of heat and cold injuries, refer to FM 21-10/Marine Corps Reference Publication (MCRP) 4-11.1D.

5-2. Heat Injuries

a. Heat injuries are environmental injuries. They may result when a service member—

- Is exposed to extreme heat, such as from the sun or from high temperatures.
- Does not wear proper clothing.
- Is in MOPP gear.
- Is inside closed spaces, such as inside an armored vehicle.
- Wears body armor.

b. Heat injury can be divided into three categories: heat cramps, heat exhaustion, and heatstroke.

c. Each service member must be able to recognize and give first aid for heat injuries.

WARNING

The heat casualty should be continually monitored for development of conditions which may require the performance of necessary basic lifesaving measures.

CAUTION

Do not use salt solutions in first aid procedures for heat injuries.

(1) Check the casualty for signs and symptoms of *cramping*.

- *Signs and symptoms.* Cramping is caused by an imbalance of chemicals (called electrolytes) in the body as a result of excessive sweating. This condition causes the casualty to exhibit:

- Cramping in the extremities (arms and legs).
- Abdominal (stomach) cramps.
- Excessive sweating.

NOTE

Thirst may or may not occur. Cramping can occur without the service member being thirsty.

- *First aid measures.*

- Move the casualty to a cool, shady area or improvise shade if none is available.

- Loosen his clothing (if not in a chemical environment).

NOTE

In a chemical environment, transport the heat casualty to a noncontaminated area as soon as the mission permits.

- Have him slowly drink at least one canteen full of water. (The body absorbs cool water faster than warm or cold water; therefore, cool water is preferred if it is available.)

- Seek medical assistance should cramps continue.

(2) Check the casualty for signs and symptoms of *heat exhaustion*.

- *Signs and symptoms.* Heat exhaustion is caused by loss of body fluids (dehydration) through sweating without adequate fluid replacement. It can occur in an otherwise fit individual who is involved in physical exertion in any hot environment especially if the service member is not acclimatized to that environment. These signs and symptoms are—

- Excessive sweating with pale, moist, cool skin.
- Headache.
- Weakness.

- Dizziness.
- Loss of appetite.
- Cramping.
- Nausea (with or without vomiting).
- Urge to defecate.
- Chills (gooseflesh).
- Rapid breathing.
- Tingling of hands and/or feet.
- Confusion.

- *First aid measures.*

- Move the casualty to a cool, shady area or improvise shade if none is available.
- Loosen or remove his clothing and boots (unless in a chemical environment); pour water on him and fan him.
- Have him slowly drink at least one canteen of water.
- Elevate his legs.
- If possible, the casualty should not participate in strenuous activity for the remainder of the day.
- Monitor the casualty until the symptoms are gone, or medical assistance arrives.

(3) Check the casualty for signs and symptoms of *heatstroke*.

WARNING

Heatstroke is a medical emergency which may result in death if care is delayed.

- *Signs and symptoms*. A service member suffering from heatstroke has been exposed to high temperatures (such as direct sunlight) or been dressed in protective overgarments, which causes the body temperature to rise. Heatstroke occurs more rapidly in service members who are engaged in work or other physical activity in a high heat environment. Heatstroke is caused by a failure of the body's cooling mechanism which includes a decrease in the body's ability to produce sweat. The casualty's skin is red (flushed), hot, and dry. He may experience weakness, dizziness, confusion, headaches, seizures, nausea, stomach pains or cramps, and his respiration and pulse may be rapid and weak. Unconsciousness and collapse may occur suddenly.

- *First aid measures*. Cool casualty immediately by—

  - Moving him to a cool, shady area or improvising shade if none is available.

  - Loosening or removing his clothing (except in a chemical environment).

  - Spraying or pouring water on him; fanning him to permit the coolant effect of evaporation.

  - Massaging his extremities and skin, which increases the blood flow to those body areas, thus aiding the cooling process.

  - Elevating his legs.

  - Having him slowly drink at least one canteen full of water if he is conscious.

NOTE

Start cooling casualty immediately. Continue cooling while awaiting transportation and during transport to an MTF.

- *Medical assistance*. Seek medical assistance because the casualty should be transported to an MTF as soon as possible. Do not interrupt the cooling process or lifesaving measures to seek help; if someone else is present send them for help. The casualty should be continually monitored for development of conditions that may require the performance of necessary basic lifesaving measures.

d. *Table*. See Table 5-1 for further information.

Table 5-1. Heat Injuries

INJURIES	SIGNS AND SYMPTOMS	FIRST AID[1]
HEAT CRAMPS	THE CASUALTY EXPERIENCES MUSCLE CRAMPS OF THE ARMS, LEGS, AND/OR STOMACH. THE CASUALTY MAY ALSO HAVE EXCESSIVE SWEATING.	1. MOVE THE CASUALTY TO A COOL SHADY AREA OR IMPROVISE SHADE AND LOOSEN CLOTHING.[2] 2. HAVE HIM SLOWLY DRINK AT LEAST ONE CANTEEN FULL OF COOL WATER SLOWLY. 3. MONITOR THE CASUALTY AND GIVE HIM MORE WATER AS TOLERATED.
HEAT EXHAUSTION	THE CASUALTY EXPERIENCES HEAVY SWEATING WITH PALE, MOIST, COOL SKIN; HEADACHE, WEAKNESS, DIZZINESS, AND/OR LOSS OF APPETITE, HEAT CRAMPS, NAUSEA (WITH OR WITHOUT VOMITING), URGE TO DEFECATE, CHILLS (GOOSE-FLESH), RAPID BREATHING, CONFUSION, AND TINGLING OF THE HANDS AND/OR FEET.	1. MOVE THE CASUALTY TO A COOL, SHADY AREA OR IMPROVISE SHADE AND LOOSEN OR REMOVE HIS CLOTHING.[2] 2. POUR WATER ON HIM AND FAN HIM TO PERMIT THE COOLANT EFFECT OF EVAPORATION. 3. HAVE HIM SLOWLY DRINK AT LEAST ONE CANTEEN FULL OF COOL WATER. 4. ELEVATE THE CASUALTY'S LEGS. 5. SEEK MEDICAL ASSISTANCE IF SYMPTOMS CONTINUE; MONITOR UNTIL SYMPTOMS ARE GONE OR MEDICAL ASSISTANCE ARRIVES.
HEATSTROKE[3] (SUNSTROKE)	THE CASUALTY STOPS SWEATING (RED [FLUSHED] HOT, DRY SKIN). HE FIRST MAY EXPERIENCE HEADACHE, DIZZINESS, NAUSEA, FAST PULSE AND RESPIRATION, SEIZURES, AND MENTAL CONFUSION. HE MAY COLLAPSE	1. MOVE THE CASUALTY TO A COOL, SHADY AREA OR IMPROVISE SHADE AND LOOSEN OR REMOVE HIS CLOTHING, REMOVE THE OUTER GARMENTS AND PROTECTIVE

Table 5-1. Heat Injuries (Continued)

INJURIES	SIGNS AND SYMPTOMS	FIRST AID[1]
	AND SUDDENLY BECOME UNCONSCIOUS. **THIS IS A MEDICAL EMERGENCY.**	CLOTHING IF THE SITUATION PERMITS.[2] 2. START COOLING THE CASUALTY IMMEDIATELY. SPRAY OR POUR WATER ON HIM. FAN HIM. MASSAGE HIS EXTREMITIES AND SKIN. 3. ELEVATE HIS LEGS. 4. IF CONSCIOUS, HAVE HIM SLOWLY DRINK AT LEAST ONE CANTEEN FULL OF COOL WATER. 5. SEEK MEDICAL AID. CONTINUE COOLING WHILE AWAITING TRANSPORT AND CONTINUE FIRST AID WHILE EN ROUTE.

LEGEND:

1 THE FIRST AID PROCEDURE FOR HEAT RELATED INJURIES CAUSED BY WEARING INDIVIDUAL PROTECTIVE EQUIPMENT (IPE) IS TO MOVE THE CASUALTY TO A CLEAN AREA AND GIVE HIM WATER TO DRINK.
2 WHEN IN A CHEMICAL ENVIRONMENT, DO NOT LOOSEN OR REMOVE THE CASUALTY'S CLOTHING.
3 CAN BE FATAL IF NOT PROVIDED FIRST AID AND MEDICAL TREATMENT PROMPTLY.

5-3. Cold Injuries

Cold injuries are most likely to occur when conditions are moderately cold, but accompanied by wet or windy conditions. Cold injuries can usually be prevented. Well-disciplined and well-trained service members can be protected even in the most adverse circumstances. They and their leaders must know the hazards of exposure to the cold. They must know the importance of personal hygiene, exercise, care of the feet and hands, and the use of protective clothing.

a. Contributing Factors.

(1) Temperature, humidity, precipitation, and wind greatly increase likelihood of cold injuries, and the service members with wet clothing are at great risk of cold injuries. Riverine operations (river, swamp, and stream crossings) increase likelihood of cold injuries. Low temperatures and low relative humidity (dry cold) promote frostbite. Higher temperatures, together with moisture, promote immersion syndrome. Windchill accelerates the loss of body heat and may aggravate cold injuries.

(2) Relatively stationary activities such as being in an observation post or on guard duty increase the service member's vulnerability to cold injury. Also, a service member is more likely to receive a cold injury if he is—

- In contact with the ground (such as marching, performing guard duty, or engaging in other outside activities).

- Immobile for long periods (such as while riding in an unheated or open vehicle).

- Standing in water, such as in a foxhole.

- Out in the cold for days without being warmed.

- Deprived of an adequate diet and rest.

- Not able to take care of his personal hygiene.

(3) Physical fatigue contributes to apathy, which leads to inactivity, personal neglect, carelessness, and reduced heat production. In turn, these increase the risk of cold injury. Service members with prior cold injuries have a higher-than-normal risk of subsequent cold injury; not necessarily involving the body part previously injured.

(4) Depressed or unresponsive service members are also vulnerable because they are less active. These service members tend to be careless about precautionary measures, especially warming activities, when cold injury is a threat.

(5) Excessive use of alcohol or drugs leading to faulty judgment or unconsciousness in a cold environment increases the risk of becoming a cold injury casualty.

b. Signs and Symptoms. Once a service member becomes familiar with the factors that contribute to cold injury, he must learn to recognize cold injury signs and symptoms.

(1) Many service members suffer cold injury without realizing what is happening to them. They may be cold and generally uncomfortable. These service members often do not notice the injured part because it is already numb from the cold.

(2) Superficial cold injury usually can be detected by numbness or tingling sensations. These signs and symptoms often can be relieved simply by loosening boots or other clothing and by exercising to improve circulation. In more advanced cases involving deep cold injury, the service member often is not aware that there is a problem until the affected part feels like a stump or block of wood.

(3) Outward signs of cold injury include discoloration of the skin at the site of injury. In light-skinned persons, the skin first reddens and then becomes pale or waxy white. In dark-skinned persons, grayness in the skin is usually evident. An injured foot or hand feels cold to the touch. Swelling may be an indication of deep injury. Also note that blisters may occur after rewarming the affected parts. Service members should work in pairs (buddy teams) to check each other for signs of discoloration and other symptoms.

c. First Aid Measures. First aid for cold injuries depends on whether they are superficial or deep. Rewarming the affected part using body heat can adequately treat cases of superficial cold injury. (For example, this can be done by covering cheeks with hands, putting fingertips in armpits, or placing the casualty's feet under the clothing of a buddy [next to his belly].) The injured part should **NOT** be massaged, exposed to a fire or stove, rubbed with snow, slapped, chafed, or soaked in cold water. Walking on injured feet should be avoided. Deep cold injury (frostbite) is very serious and requires prompt first aid to avoid or to minimize the loss of parts or all of the fingers, toes, hands, or feet. The sequence for treating cold injuries depends on whether the condition is life-threatening. The first priority in managing cold injuries is to remove the casualty from the cold environment (such as building an improvised shelter). Other injuries the casualty may have are provided first aid simultaneously while waiting for transportation or evacuation. If the casualty is to be transported in a nonmedical vehicle, first aid measures should be continued en route to the MTF.

d. Conditions Caused by Cold. Conditions caused by cold include chilblain, immersion syndrome (immersion foot and trench foot), frostbite, snow blindness, dehydration, and hypothermia.

(1) *Chilblain.*

- *Signs and symptoms.* Chilblain is caused by repeated prolonged exposure of bare skin at temperatures from 60° Fahrenheit (F) to

32°F, or 20°F for acclimated, dry, unwashed skin. The area may be acutely swollen, red, tender, and hot with itchy skin. There may be no loss of skin tissue in untreated cases but continued exposure may lead to infected, ulcerated, or bleeding lesions.

- *First aid measures.* Within minutes, the area usually responds to locally applied body heat. Rewarm the affected part by applying firm steady pressure with your hands, or placing the affected part under your arms or against the stomach of a buddy. **DO NOT** rub or massage affected areas.

NOTE

Medical personnel should evaluate the injury, because signs and symptoms of tissue damage may be slow to appear.

(2) *Immersion syndrome (immersion foot and trench foot).* Immersion foot and trench foot are injuries that result from fairly long exposure of the feet to wet conditions at temperatures from approximately 32°F to 50°F. Inactive feet in damp or wet socks and boots, or tightly laced boots which impair circulation, are even more susceptible to injury. This injury can be very serious; it can lead to loss of toes or parts of the feet. If exposure of the feet has been prolonged and severe, the feet may swell so much that pressure closes the blood vessels and cuts off circulation. Should an immersion injury occur, dry the feet thoroughly and transport the casualty to an MTF by the fastest means possible.

- *Signs and symptoms.* At first, the parts of the affected foot are cold and painless, the pulse is weak, and numbness may be present. Second, the parts may feel hot, and burning and shooting pains may begin. In later stages, the skin is pale with a bluish cast and the pulse decreases. Other signs and symptoms that may follow are blistering, swelling, redness, heat, hemorrhaging (bleeding), and gangrene.

- *First aid measures.* First aid measures are required for all stages of immersion syndrome injury. Rewarm the injured part gradually by exposing it to warm air. Protect it from trauma and secondary infections. Dry, loose clothing or several layers of warm coverings are preferable to extreme heat. Under no circumstances should the injured part be exposed to an open fire. Elevate the injured part to relieve the swelling. Transport the casualty to an MTF as soon as possible. When the part is rewarmed, the casualty often feels a burning sensation and pain. Symptoms may persist for days or weeks even after rewarming.

NOTE

When providing first aid for immersion foot and trench foot—**DO NOT** massage the injured part. **DO NOT** moisten the skin. **DO NOT** apply heat or ice.

(3) *Frostbite*. Frostbite is the injury of tissue caused from exposure to cold, usually below 32°F depending on the windchill factor, duration of exposure, and adequacy of protection. Individuals with a history of cold injury are likely to suffer an additional cold injury. The body parts most easily frostbitten are the cheeks, nose, ears, chin, forehead, wrists, hands, and feet. Frostbite may involve only the skin (superficial), or it may extend to a depth below the skin (deep). Deep frostbite is very serious and requires prompt first aid to avoid or to minimize the loss of parts or all of the fingers, toes, hands, or feet.

- *Signs and symptoms*.
  - Loss of sensation (numb feeling) in any part of the body.
  - Sudden blanching (whitening) of the skin of the affected part, followed by a momentary tingling sensation.
  - Redness of skin in light-skinned service members; grayish coloring in dark-skinned service members.
  - Blisters.
  - Swelling or tender areas.
  - Loss of previous sensation of pain in affected area.
  - Pale, yellowish, waxy-looking skin.
  - Frozen tissue that feels solid (or wooden) to the touch.

CAUTION

Deep frostbite is a very serious injury and requires immediate first aid and subsequent medical treatment to avoid or minimize loss of body parts.

- *First aid measures.*

  - *Face, ears, and nose.* Cover the casualty's affected area with his and/or your bare hands until sensation and color return.

  - *Hands.* Open the casualty's field jacket and shirt. (In a chemical environment, do not loosen or remove the clothing and protective overgarments.) Place the affected hands under the casualty's armpits. Close the field jacket and shirt to prevent additional exposure.

  - *Feet.* Remove the casualty's boots and socks if he does not need to walk any further to receive additional treatment. (Thawing the casualty's feet and forcing him to walk on them will cause additional pain and injury.) Place the affected feet under clothing and against the body of another service member.

WARNING

DO NOT attempt to thaw the casualty's feet or other frozen areas if he will be required to walk or travel to an MTF for additional medical treatment. The possibility of additional injury from walking is less when the feet are frozen than when they are thawed. (However, if possible avoid walking.) Thawing in the field increases the possibilities of infection, gangrene, or other injury.

NOTE

Thawing may occur spontaneously during transportation to the MTF; this cannot be avoided since the body in general must be kept warm.

In all of the above areas, ensure that the casualty is kept warm and that he is covered (to avoid further injury). Seek medical treatment as soon as possible. Reassure the casualty, protect the affected area from further injury by covering it lightly with a blanket or any dry clothing, and seek shelter out of the wind. Remove or loosen constricting clothing (except in a contaminated environment) and increase insulation. Ensure the casualty exercises as much as possible, avoiding trauma to the injured part, and is prepared for pain when thawing occurs. Protect the frostbitten part from additional injury. **DO NOT**—

- Rub the injured part with snow or apply cold water soaks.

- Warm the part by massage or exposure to open fire because the frozen part may be burned due to the lack of feeling.

- Use ointments or other salves.

- Manipulate the part in any way to increase circulation.

- Use alcohol or tobacco because this reduces the body's resistance to cold.

NOTE

Remember, when freezing extends to a depth below the skin, it is a much more serious injury. Extra care is required to reduce or avoid the chances of losing all or part of the toes or feet. This also applies to the fingers and hands.

(4) *Snow blindness*. Snow blindness is the effect that glare from an ice field or snowfield has on the eyes. It is more likely to occur in hazy, cloudy weather than when the sun is shining. Glare from the sun will cause an individual to instinctively protect his eyes. However, in cloudy weather, he may be overconfident and expose his eyes longer than when the threat is more obvious. He may also neglect precautions such as the use of protective eyewear. Waiting until discomfort (pain) is felt before using protective eyewear is dangerous because a deep burn of the eyes may already have occurred.

- *Signs and symptoms*. Symptoms of snow blindness are a sensation of grit in the eyes with pain in and over the eyes, made worse by moving the eyeball. Other signs and symptoms are watering, redness, headache, and increased pain on exposure to light.

- *First aid measures*. First aid measures consist of blindfolding or covering the eyes with a dark cloth which stops painful eye movement. Complete rest is desirable. If further exposure to light is not preventable, the eyes should be protected with dark bandages or the darkest glasses available. Once unprotected exposure to sunlight stops, the condition usually heals in a few days without permanent damage. The casualty should be evacuated to the nearest MTF.

(5) *Dehydration*. Dehydration occurs when the body loses too much fluid. A certain amount of body fluid is lost through normal body processes. A normal daily intake of liquids replaces these losses. When individuals are engaged in any strenuous exercises or activities, fluid is lost

through sweating and this loss creates an imbalance of fluids in the body, and if not matched by rehydration it can contribute to dehydration. The danger of dehydration is as prevalent in cold regions as it is in hot regions. In hot weather, the individual is aware of his body losing fluids through sweat. In cold weather, however, it is extremely difficult to realize that this condition exists since sweating is not as apparent as in a hot environment. The danger of dehydration in cold weather operations is a serious problem. In cold climates, sweat evaporates so rapidly or is absorbed so thoroughly by layers of heavy clothing that it is rarely visible on the skin. Dehydration also occurs during cold weather operations because drinking is inconvenient. Dehydration will weaken or incapacitate a casualty for a few hours, or sometimes, several days. Because rest is an important part of the recovery process, casualties must take care that limited movement during their recuperative period does not enhance the risk of becoming a cold injury casualty.

- *Signs and symptoms.* The symptoms of cold weather dehydration are similar to those encountered in heat exhaustion. The mouth, tongue, and throat become parched and dry, and swallowing becomes difficult. The casualty may have nausea (with or without vomiting) along with extreme dizziness and fainting. The casualty may also feel generally tired and weak and may experience muscle cramps. Focusing the eyes may also become difficult.

- *First aid measures.* The casualty should be kept warm and his clothes should be loosened (if not in a chemical environment) to allow proper circulation. Shelter from wind and cold must be provided. Fluid replacement should begin immediately and the service member transported to an MTF as soon as possible.

(6) *Hypothermia (general cooling).* When exposed to prolonged cold weather a service member may become both mentally and physically numb, thus neglecting essential tasks or requiring more time and effort to achieve them. Under some conditions (particularly cold water immersion), even a service member in excellent physical condition may die in a matter of minutes. The destructive influence of cold on the body is called *hypothermia.* This means bodies lose heat faster than they can produce it. Hypothermia can occur from exposure to temperatures either above or below freezing, especially from immersion in cold water, wet-cold conditions, or from the effect of wind. Physical exhaustion and insufficient food intake may also increase the risk of hypothermia. General cooling of the entire body to a temperature below 95°F is caused by continued exposure to low or rapidly dropping temperatures, cold moisture, snow, or ice. Fatigue, poor physical condition, dehydration, faulty blood circulation, alcohol or other drug use, trauma, and immersion can cause hypothermia. Remember, cold

may affect the body systems slowly and almost without notice. Service members exposed to low temperatures for extended periods may suffer ill effects even if they are well protected by clothing.

- *Signs and symptoms*. As the body cools, there are several stages of progressive discomfort and impairment. A sign that is noticed immediately is shivering. Shivering is an attempt by the body to generate heat. The pulse is faint or very difficult to detect. People with temperatures around 90°F may be drowsy and mentally slow. Their ability to move may be hampered, stiff, and uncoordinated, but they may be able to function minimally. Their speech may be slurred. As the body temperature drops further, shock becomes evident as the person's eyes assume a glassy state, breathing becomes slow and shallow, and the pulse becomes weaker or absent. The person becomes very stiff and uncoordinated. Unconsciousness may follow quickly. As the body temperature drops even lower, the extremities freeze, and a deep (or core) body temperature (below 85°F) increases the risk of irregular heart action. This irregular heart action or heart standstill can result in sudden death.

- *First aid measures*. Except in cases of the most severe hypothermia (marked by coma or unconsciousness and a weak pulse), first aid measures for hypothermia are directed towards protecting the casualty from further loss of body heat. For the casualty who is conscious, first aid measures are directed at rewarming the body evenly and without delay. Provide heat by using a hot water bottle or field expedient or another service member's body heat.

CAUTION

DO NOT expose the casualty to an open fire, as he may become burned.

NOTE

When using a hot water bottle or field expedient (canteen filled with warm water), the bottle or canteen must be wrapped in cloth prior to placing it next to the casualty. This will reduce the chance of burning the casualty's skin.

Always call or send for help as soon as possible and protect the casualty immediately with dry clothing or a sleeping bag. Then, move him to a warm place. Evaluate other injuries and provide first aid as required. First aid measures can be performed while the casualty is waiting transportation or while he is en route. In the case of an accidental breakthrough into ice water, or other hypothermic accident, strip the casualty of wet clothing immediately and bundle him into a sleeping bag. Rescue breathing should be started at once if the casualty's breathing has stopped or is irregular or shallow. Warm liquids (**NOT HOT**) may be given gradually if the casualty is conscious. **DO NOT** force liquids on an unconscious or semiconscious casualty because he may choke. The casualty should be transported on a litter because the exertion of walking may aggravate circulation problems. Medical personnel should immediately treat any hypothermia casualty. Hypothermia is life threatening until normal body temperature has been restored. The first aid measures for a casualty with severe hypothermia are based upon the following principles: attempt to avoid further heat loss, handle the casualty gently, and transport the casualty as soon as possible to the nearest MTF. If at all possible, the casualty should be evacuated by medical personnel.

WARNING

Rewarming a severely hypothermic casualty is extremely dangerous in the field due to the possibility of such complications as rewarming, shock and disturbances in the rhythm of the heartbeat. These conditions require treatment by medical personnel.

NOTE

Resuscitation of casualties with hypothermic complications is difficult if not impossible to do outside of an MTF setting.

CAUTION

The casualty is unable to generate his own body heat. Therefore, merely placing him in a blanket or sleeping bag is not sufficient.

e. *Table.* See Table 5-2 for further information.

Table 5-2. Injuries Caused by Cold and Wet Conditions

INJURIES	SIGNS/SYMPTOMS	FIRST AID
CHILBLAIN	RED SWOLLEN, HOT, TENDER, ITCHING SKIN. CONTINUED EXPOSURE MAY LEAD TO INFECTED (ULCERATED OR BLEEDING) SKIN LESIONS.	1. AREA USUALLY RESPONDS TO LOCALLY APPLIED REWARMING (BODY HEAT). 2. DO NOT RUB OR MASSAGE AREA. 3. SEEK MEDICAL AID.
IMMERSION SYNDROME (IMMERSION FOOT/TRENCH FOOT)	AFFECTED PARTS ARE COLD, NUMB, AND PAINLESS. PARTS MAY THEN BE HOT, WITH BURNING AND SHOOTING PAINS. ADVANCED STAGE: SKIN PALE WITH BLUISH CAST; PULSE DECREASES; BLISTERING, SWELLING, HEAT, HEMORRHAGING, AND GANGRENE MAY FOLLOW.	1. GRADUAL REWARMING BY EXPOSURE TO WARM AIR. 2. DO NOT MASSAGE OR MOISTEN SKIN. 3. PROTECT AFFECTED PARTS FROM TRAUMA. 4. DRY FEET THOROUGHLY, AVOID WALKING. 5. SEEK MEDICAL AID.
FROSTBITE	LOSS OF SENSATION (NUMB FEELING) IN ANY PART OF THE BODY. SUDDEN BLANCHING (WHITENING) OF THE SKIN OF THE AFFECTED PART, FOLLOWED BY A MOMENTARY TINGLING SENSATION. REDNESS OF SKIN IN LIGHT-SKINNED SERVICE MEMBERS; GRAYISH COLORING IN DARK-SKINNED SERVICE MEMBERS. BLISTERS. SWELLING OR TENDER AREAS. LOSS OF PREVIOUS SENSATION OF PAIN IN THE AFFECTED AREA. PALE YELLOWISH, WAXY-LOOKING SKIN. FROZEN TISSUE THAT FEELS SOLID (WOODEN) TO THE TOUCH.	1. WARM THE AREA AT THE FIRST SIGN OF FROSTBITE, USING FIRM, STEADY PRESSURE OF THE HAND, UNDERARM, OR ABDOMEN. 2. FACE, EARS, NOSE: COVER AREA WITH HANDS (CASUALTY'S OWN OR BUDDY'S). 3. HANDS: OPEN FIELD JACKET AND PLACE CASUALTY'S HANDS AGAINST HIS BODY, THEN CLOSE THE JACKET TO PREVENT HEAT LOSS. 4. FEET: REMOVE THE CASUALTY'S BOOTS AND SOCKS AND PLACE HIS FEET AGAINST THE BODY OF ANOTHER SERVICE MEMBER.

Table 5-2. Injuries Caused by Cold and Wet Conditions (Continued)

INJURIES	SIGNS/SYMPTOMS	FIRST AID
		5. WARNING: DO NOT ATTEMPT TO THAW THE CASUALTY'S FEET OR OTHER FROZEN AREAS IF HE WILL BE REQUIRED TO WALK OR TRAVEL TO AN MTF FOR ADDITIONAL TREATMENT. THE POSSIBILITY OF INJURY FROM WALKING IS LESS WHEN THE FEET ARE FROZEN THAN WHEN THEY HAVE BEEN THAWED. (HOWEVER, IF POSSIBLE AVOID WALKING.) THAWING IN THE FIELD INCREASES THE POSSIBILITY OF INFECTION, GANGRENE, OR INJURY. 6. LOOSEN OR REMOVE CONSTRICTING CLOTHING AND REMOVE ANY JEWELRY. 7. INCREASE INSULATION (COVER WITH BLANKET OR OTHER DRY MATERIAL). ENSURE CASUALTY EXERCISES AS MUCH AS POSSIBLE, AVOIDING TRAUMA TO INJURED PART.
SNOW BLINDNESS	EYES MAY FEEL SCRATCHY. WATERING, REDNESS, HEADACHE, AND INCREASED PAIN WITH EXPOSURE TO LIGHT CAN OCCUR.	1. COVER THE EYES WITH A DARK CLOTH. 2. SEEK MEDICAL AID.
DEHYDRATION	SIMILAR TO HEAT EXHAUSTION (REFER TO TABLE 5-1).	1. KEEP WARM. 2. CASUALTY NEEDS FLUID REPLACEMENT, REST, AND PROMPT MEDICAL AID.
HYPOTHERMIA	CASUALTY IS COLD. SHIVERING. CORE TEMPERATURE IS LOW. CONSCIOUSNESS MAY BE	**MILD HYPOTHERMIA** 1. REWARM BODY

Table 5-2. Injuries Caused by Cold and Wet Conditions (Continued)

INJURIES	SIGNS/SYMPTOMS	FIRST AID
	ALTERED. UNCOORDINATED MOVEMENTS MAY OCCUR. SHOCK AND COMA MAY RESULT AS BODY TEMPERATURE DROPS.	EVENLY AND WITHOUT DELAY. (NEED TO PROVIDE HEAT SOURCE; CASUALTY'S BODY UNABLE TO GENERATE HEAT.) 2. KEEP DRY, PROTECT FROM THE ELEMENTS. 3. WARM (NOT HOT) LIQUIDS MAY BE GIVEN GRADUALLY (TO CONSCIOUS CASUALTIES ONLY). 4. BE PREPARED TO START BASIC LIFE SUPPORT MEASURES FOR THE CASUALTY. 5. SEEK MEDICAL TREATMENT IMMEDIATELY. **SEVERE HYPOTHERMIA** 1. STABILIZE THE TEMPERATURE. 2. ATTEMPT TO AVOID FURTHER HEAT LOSS. 3. HANDLE THE CASUALTY GENTLY. 4. EVACUATE TO THE NEAREST MTF AS SOON AS POSSIBLE. **5. WARNING: HYPOTHERMIA IS A MEDICAL EMERGENCY. PROMPT MEDICAL ATTENTION IS NECESSARY.**

CHAPTER 6

FIRST AID FOR BITES AND STINGS

6-1. General

Snakebites, insect bites, or stings can cause intense pain and/or swelling. If not treated promptly and correctly, they can cause serious illness or death. The severity of a snakebite depends upon: whether the snake is poisonous or nonpoisonous, the type of snake, the location of the bite, and the amount of venom injected. Bites from humans and other animals, such as dogs, cats, bats, raccoons, and rats, can cause severe bruises and infection and tears or lacerations of tissue. Awareness of the potential sources of injuries can reduce or prevent them from occurring. Knowledge and prompt application of first-aid measures can lessen the severity of injuries from bites and stings and keep the service member from becoming a serious casualty.

6-2. Types of Snakes

a. Nonpoisonous Snakes. There are approximately 130 different varieties of nonpoisonous snakes in the United States. They have oval-shaped heads and round eyes. Unlike poisonous snakes, discussed below, nonpoisonous snakes do not have fangs with which to inject venom. Figure 6-1 depicts the characteristics of a nonpoisonous snake.

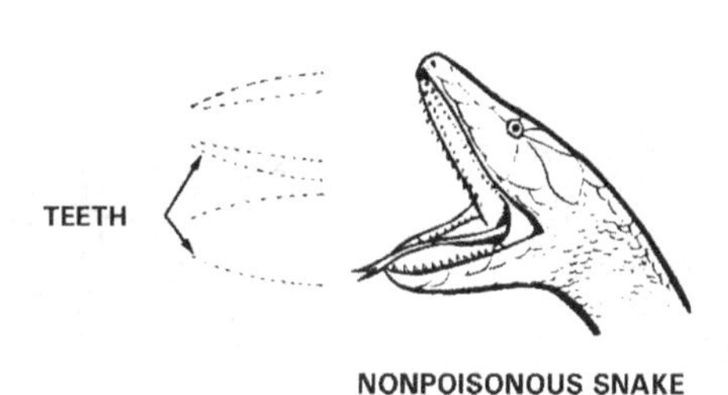

Figure 6-1. Characteristics of nonpoisonous snake.

b. Poisonous Snakes. Poisonous snakes are found throughout the world, primarily in tropical to moderate climates. Within the United States, there are four kinds: rattlesnakes, copperheads, water moccasins (cottonmouth), and coral snakes. Poisonous snakes in other parts of the world include sea snakes, the fer-de-lance, the bushmaster, and the tropical rattlesnake in tropical Central America; the Malayan pit viper in the tropical Far East; the cobra in Africa and Asia; the mamba (or black mamba) in central and southern Africa; and the krait in India and Southeast Asia. Refer to Figure 6-2 for characteristics of a poisonous pit viper.

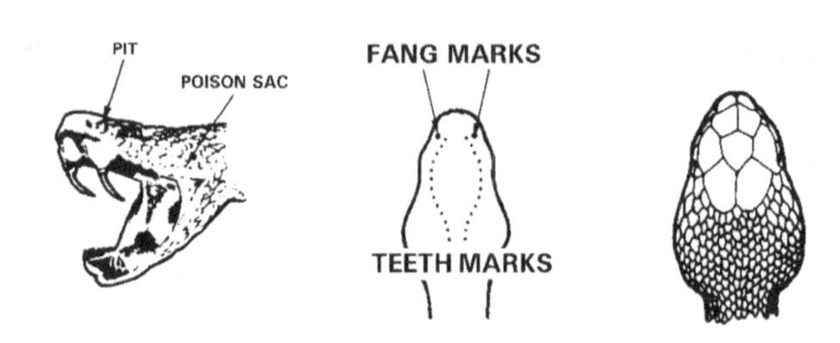

Figure 6-2. Characteristics of poisonous pit viper.

c. Pit Vipers (Poisonous). Figure 6-3 depicts a variety of poisonous snakes.

Figure 6-3. Poisonous snakes.

(1) Rattlesnakes, bushmasters, copperheads, fer-de-lance, Malayan pit vipers, and water moccasins (cottonmouth) are called pit vipers because of the small, deep pits between the nostrils and eyes on each side of the head (Figure 6-2). In addition to their long, hollow fangs, these snakes have other identifying features: thick bodies, slit-like pupils of the eyes, and flat, almost triangular-shaped heads. Color markings and other identifying characteristics, such as rattles or a noticeable white interior of the mouth (cottonmouth), also help distinguish these poisonous snakes. Further

identification is provided by examining the bite pattern of the wound for signs of fang entry. Occasionally there will be only one fang mark, as in the case of a bite on a finger or toe where there is no room for both fangs, or when the snake has broken off a fang.

(2) The casualty's condition provides the best information about the seriousness of the situation, or how much time has passed since the bite occurred. Pit viper bites are characterized by severe burning pain. Discoloration and swelling around the fang marks usually begins within 5 to 10 minutes after the bite. If only minimal swelling occurs within 30 minutes, the bite will almost certainly have been from a nonpoisonous snake or possibly from a poisonous snake which did not inject venom. The venom destroys blood cells, causing a general discoloration of the skin. Blisters and numbness in the affected area follow this reaction. Other signs, which can occur, are weakness, rapid pulse, nausea, shortness of breath, vomiting, and shock.

d. Corals, Cobras, Kraits, and Mambas. Corals (Figure 6-4), cobras (Figure 6-5), kraits, and mambas all belong to the same group even though they are found in different parts of the world. All four inject their venom through short, grooved fangs, leaving a characteristic bite pattern.

Figure 6-4. Coral snake.

(1) The small coral snake, found in the Southeastern US, is brightly colored with bands of red, yellow (or almost white), and black completely encircling the body. Other nonpoisonous snakes have the same coloring, but on the coral snake found in the US, the red ring always touches the yellow ring. To know the difference between a harmless snake and the coral snake found in the United States, remember the following:

"Red on yellow will kill a fellow,
Red on black, venom will lack."

Figure 6-5. Cobra snake.

(2) The venom of corals, cobras, kraits, and mambas produces symptoms different from those of pit vipers. Because there is only minimal pain and swelling, many people believe that the bite is not serious. Delayed reactions in the nervous system normally occur between 1 to 7 hours after the bite. Symptoms include blurred vision, drooping eyelids, slurred speech, drowsiness, and increased salivation and sweating. Nausea, vomiting, shock, respiratory difficulty, paralysis, convulsions, and coma will usually develop if the bite is not treated promptly.

e. Sea Snakes. Sea snakes (Figure 6-6) are found in the warm water areas of the Pacific and Indian oceans, along the coasts, and at the mouths of some larger rivers. Their venom is **VERY** poisonous, but their fangs are only 1/4 inch long. The first aid outlined for land snakes also applies to sea snakes.

Figure 6-6. Sea snake.

6-3. Snakebites

a. Poisonous snakes **DO NOT** always inject venom when they bite or strike a person. However, all snakes may carry tetanus (lockjaw); anyone bitten by a snake, whether poisonous or nonpoisonous, should immediately seek medical attention.

- Poison is injected from the venom sacs through grooved or hollow fangs. Depending on the species, these fangs are either long or short. Pit vipers have long hollow fangs. These fangs are folded against the roof of the mouth and extend when the snake strikes. This allows them to strike quickly and then withdraw. Cobras, coral snakes, kraits, mambas, and sea snakes have short, grooved fangs. These snakes are less effective in their attempts to bite, since they must chew after striking to inject enough venom (poison) to be effective. Figure 6-7 depicts the characteristics of a poisonous snakebite.

- In the event you are bitten, attempt to identify and/or kill the snake. Take it to medical personnel for inspection/identification. This provides valuable information to medical personnel who deal with snakebites. **TREAT ALL SNAKEBITES AS POISONOUS**.

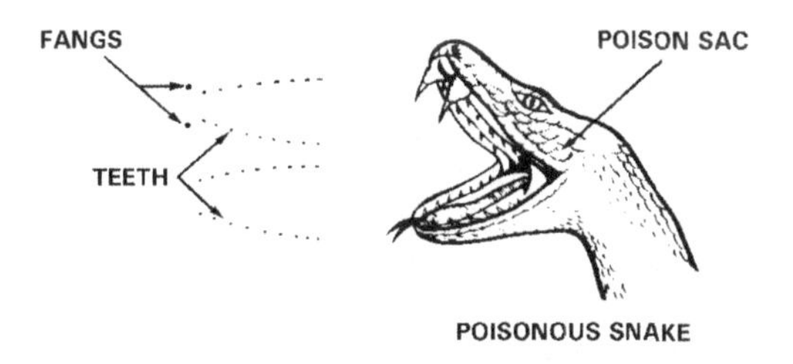

Figure 6-7. Characteristics of poisonous snakebite.

b. The venoms of different snakes cause different effects. Pit viper venom (hemotoxin [blood toxin]) destroys tissue and blood cells. Cobras, adders, and coral snakes inject powerful venom (neurotoxin [nerve toxin]) which affect the central nervous system, causing respiratory paralysis. Water moccasins and sea snakes have venom that is both hemotoxic and neurotoxic.

c. The identification of poisonous snakes is very important since medical treatment will be different for each type of venom. *Unless it can be positively identified, the snake should be killed and saved.* When this is not possible or when doing so is a serious threat to others, identification may sometimes be difficult since many venomous snakes resemble harmless varieties. When dealing with snakebite problems in foreign countries, seek advice, professional or otherwise, which may help identify species in the particular area of operations.

d. Get the casualty to an MTF as soon as possible and with minimum movement. Until evacuation or treatment is possible, have the casualty lie quietly and not move any more than necessary. If the casualty has been bitten on an extremity, **DO NOT** elevate the limb; keep the extremity level with the body. Keep the casualty comfortable and reassure him. If the casualty is alone when bitten, he should go to the medical facility himself rather than wait for someone to find him. Unless the snake has been positively identified, attempt to kill it and send it with the casualty. Be sure that retrieving the snake does not endanger anyone or delay transporting the casualty.

(1) If the bite is on an arm or leg, place a constricting band (narrow cravat [swathe], or narrow gauze bandage) one to two fingerbreadths above and below the bite (Figure 6-8). If the bite is on the hand or foot, place a single band above the wrist or ankle. The band should be tight enough to stop the flow of blood near the skin, but not tight enough to interfere with circulation. In other words, it should not have a tourniquet-like affect. If no swelling is seen, place the bands about 1 inch from either side of the bite. If swelling is present, put the bands on the unswollen part at the edge of the swelling. If the swelling extends beyond the band, move the band to the new edge of the swelling. (If possible, leave the old band on, place a new one at the new edge of the swelling, and then remove and save the old one in case the process has to be repeated.)

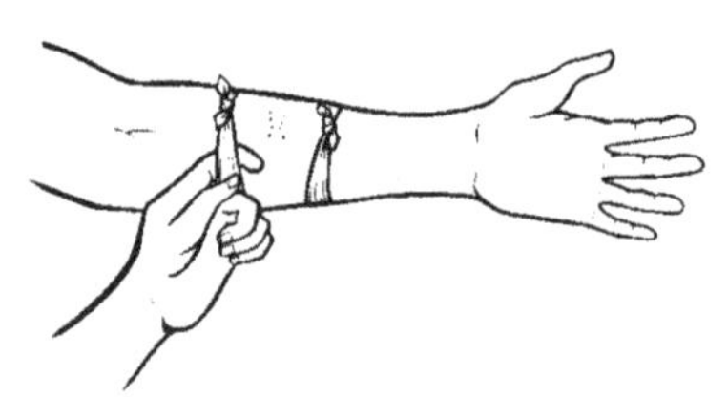

Figure 6-8. Constricting band.

CAUTION

DO NOT attempt to cut open the bite nor suck out the venom. If the venom should seep through any damaged or lacerated tissues in your mouth, you could immediately lose consciousness or even die.

(2) If the bite is located on an arm or leg, immobilize it at a level below the heart. **DO NOT** elevate an arm or leg even with or above the level of the heart.

CAUTION

When a splint is used to immobilize the arm or leg, take **EXTREME** care to ensure the splinting is done properly and does not bind. Watch it closely and adjust it if any changes in swelling occur.

(3) When possible, clean the area of the bite with soap and water. **DO NOT** use ointments of any kind.

(4) **NEVER** give the casualty food, alcohol, stimulants (coffee or tea), drugs, or tobacco.

(5) Remove rings, watches, or other jewelry from the affected limb.

6-4. Human or Animal Bites

Human or other land animal bites may cause lacerations or bruises. In addition to damaging tissue, bites always present the possibility of infection.

a. Human Bites. Human bites that break the skin may become seriously infected since the mouth is heavily contaminated with bacteria. Medical personnel **MUST** treat all human bites.

b. Animal Bites. Land animal bites can result in both infection and disease. Tetanus, rabies, and various types of fevers can follow an untreated animal bite. Because of these possible complications, the animal causing the bite should, if possible, be captured or killed (without damaging its head) so that it can be tested for disease.

c. First Aid.

(1) Cleanse the wound thoroughly with soap.

(2) Flush it well with water.

(3) Cover it with a sterile dressing.

(4) Immobilize the injured arm or leg, if appropriate.

(5) Transport the casualty immediately to an MTF.

NOTE

If unable to capture or kill the animal, provide medical personnel with any information that will help identify it.

6-5. Marine (Sea) Animals

With the exception of sharks and barracuda, most marine animals will not deliberately attack. The most frequent injuries from marine animals are wounds by biting, stinging, or puncturing. Wounds inflicted by marine animals can be very painful, but are rarely fatal.

a. Sharks, Barracuda, and Alligators. Wounds from these marine animals can involve major trauma as a result of bites and lacerations. Bites from large marine animals are potentially the most life threatening of all injuries from marine animals. Major wounds from these animals can be treated by controlling the bleeding, preventing shock, giving basic life support, splinting the injury, and by securing prompt medical aid.

b. Turtles, Moray Eels, and Corals. These animals normally inflict minor wounds. Treat by cleansing the wound(s) thoroughly and by splinting if necessary.

c. Jellyfish, Portuguese Man-of-War, Anemones, and Others. This group of marine animals inflict injury by means of stinging cells in their tentacles. Contact with the tentacles produces burning pain with a rash and small hemorrhages on the skin. Shock, muscular cramping, nausea, vomiting, and respiratory distress may also occur. Gently remove the clinging tentacles with a towel and wash or treat the area. Use diluted ammonia or alcohol, meat tenderizer, and talcum powder. If symptoms become severe or persist, seek medical assistance.

d. Spiny Fish, Urchins, Stingrays, and Cone Shells. These animals inject their venom by puncturing the skin with their spines. General signs and symptoms include swelling, nausea, vomiting, generalized cramps, diarrhea, muscular paralysis, and shock. Deaths are rare. Treatment consists of soaking the wounds in hot water (when available) for 30 to 60 minutes. This inactivates the heat sensitive toxin. In addition, further first aid measures (controlling bleeding, applying a dressing, and so forth) should be carried out as necessary.

CAUTION

Be careful not to scald the casualty with water that is too hot because the pain of the wound will mask the normal reaction to heat.

6-6. Insect (Arthropod) Bites and Stings

An insect bite or sting can cause great pain, allergic reaction, inflammation, and infection. If not treated correctly, some bites/stings may cause serious illness or even death. When an allergic reaction is not involved, first aid is a simple process. In any case, medical personnel should examine the casualty at the earliest possible time. It is important to properly identify the spider, bee, or creature that caused the bite/sting, especially in cases of allergic reaction.

a. Types of Insects. The insects found throughout the world that can produce a bite or sting are too numerous to mention in detail. Commonly encountered stinging or biting insects include brown recluse spiders (Figure 6-9), black widow spiders (Figure 6-10), tarantulas (Figure 6-11), scorpions (Figure 6-12), urticating caterpillars, bees, wasps, centipedes, conenose beetles (kissing bugs), ants, and wheel bugs. Upon being reassigned, especially to overseas areas, take the time to become acquainted with the types of insects to avoid.

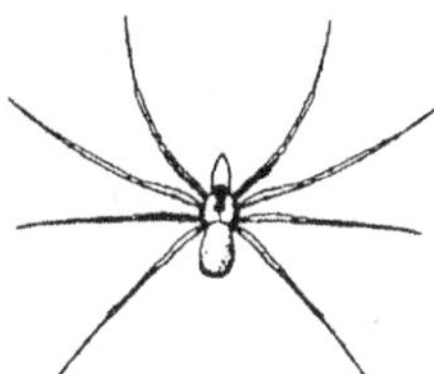

Figure 6-9. Brown recluse spider.

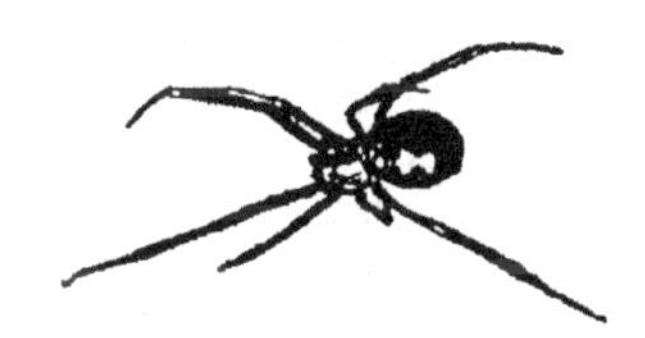

Figure 6-10. Black widow spider.

Figure 6-11. Tarantula.

Figure 6-12. Scorpion.

b. Signs and Symptoms. Discussed in paragraphs (1) and (2) below are the most common effects of insect bites/stings. They can occur alone or in combination with the others.

(1) *Less serious.* Commonly seen signs/symptoms are pain, irritation, swelling, heat, redness, and itching. Hives or wheals (raised areas of the skin that itch) may occur. These are the least severe of the allergic reactions that commonly occur from insect bites/stings. They are usually dangerous only if they affect the air passages (mouth, throat, nose, and so forth), which could interfere with breathing. The bites/stings of bees, wasps, ants, mosquitoes, fleas, and ticks are usually not serious and normally produce mild and localized symptoms. A tarantula's bite is usually no worse than that of a bee sting. Scorpions are rare and their stings (except for a specific species found only in the Southwest desert) are painful but usually not dangerous.

(2) *Serious.* Emergency allergic or hypersensitive reactions sometimes result from the stings of bees, wasps, and ants. Many people are allergic to the venom of these particular insects. Bites or stings from these insects may produce more serious reactions, to include generalized itching and hives, weakness, anxiety, headache, breathing difficulties, nausea, vomiting, and diarrhea. Very serious allergic reactions (called *anaphylactic shock*) can lead to complete collapse, shock, and even death. Spider bites (particularly from the black widow and brown recluse spiders) can also be

serious. Venom from the black widow spider affects the nervous system. This venom can cause muscle cramps, a rigid, nontender abdomen, breathing difficulties, sweating, nausea, and vomiting. The brown recluse spider generally produces local rather than system-wide problems; however, local tissue damage around the bite can be severe and can lead to an ulcer and even gangrene.

c. *First Aid.* There are certain principles that apply regardless of what caused the bite/sting. Some of these are—

- If there is a stinger present (for example, from a bee), remove the stinger by scraping the skin's surface with a fingernail or knife. **DO NOT** squeeze the sac attached to the stinger because it may inject more venom.

- Wash the area of the bite/sting with soap and water (alcohol or an antiseptic may also be used) to help reduce the chances of an infection and remove traces of venom.

- Remove jewelry from bitten extremities because swelling may occur.

- In most cases of insect bites the reaction will be mild and localized; use ice or cold compresses (if available) on the site of the bite/sting. This will help reduce swelling, ease the pain, and slow the absorption of venom. Meat tenderizer (to neutralize the venom) or calamine lotion (to reduce itching) may be applied locally. If necessary, seek medical assistance.

- In more serious reactions (severe and rapid swelling, allergic symptoms, and so forth) treat the bite/sting like you would treat a snakebite; that is, apply constricting bands above and below the site.

- Be prepared to perform basic life-support measures, such as rescue breathing.

- Reassure the casualty and keep him calm.

- In serious reactions, attempt to capture the insect for positive identification; however, be careful not to become a casualty yourself.

- If the reaction to the bite/sting appears serious, seek medical assistance.

WARNING

Insect bites/stings may cause *anaphylactic shock* (a shock caused by a severe allergic reaction). This is a life-threatening event and a TRUE MEDICAL EMERGENCY. Be prepared to perform the basic life-support measures and to immediately transport the casualty to an MTF.

NOTE

Be aware that some allergic or hypersensitive individuals may carry identification or emergency insect bite treatment kits. If the casualty is having an allergic reaction and has such a kit, administer the medication in the kit according to the instructions which accompany the kit.

d. Supplemental Information. For additional information concerning biting insects, see FM 21-10.

6-7. First Aid for Bites and Stings

See the table below for information on bites and stings.

Table 6-1. First Aid Measures for Bites and Stings

TYPES	FIRST AID MEASURES
SNAKEBITE	1. MOVE CASUALTY AWAY FROM THE SNAKE. 2. REMOVE JEWELRY FROM THE AFFECTED AREA, IF APPLICABLE. 3. REASSURE CASUALTY AND KEEP HIM QUIET. 4. APPLY CONSTRICTING BAND, 1-2 FINGERBREADTHS FROM THE BITE. YOU SHOULD BE ABLE TO INSERT A FINGER BETWEEN THE BAND AND THE SKIN. a. *ARM OR LEG BITE.* PLACE ONE BAND ABOVE AND ONE BAND BELOW THE BITE SITE. b. *HAND OR FOOT BITE.* PLACE ONE BAND ABOVE THE WRIST OR ANKLE.

Table 6-1. First Aid Measures for Bites and Stings

TYPES	FIRST AID MEASURES
	5. IMMOBILIZE THE AFFECTED PART IN A POSITION BELOW THE LEVEL OF THE HEART. 6. KILL THE SNAKE (IF POSSIBLE, WITHOUT DAMAGING ITS HEAD OR ENDANGERING YOURSELF) AND SEND IT WITH THE CASUALTY. 7. SEEK MEDICAL ASSISTANCE IMMEDIATELY.
BROWN RECLUSE SPIDER OR BLACK WIDOW SPIDER BITE	1. KEEP CASUALTY QUIET. 2. REMOVE ALL JEWELRY FROM AFFECTED PART, IF APPLICABLE. 3. WASH THE AREA. 4. APPLY ICE OR FREEZE PACK, IF AVAILABLE. 5. SEEK MEDICAL ASSISTANCE.
TARANTULA BITE OR SCORPION STING OR ANT BITE	1. WASH THE AREA. 2. REMOVE ALL JEWELRY FROM AFFECTED PART, IF APPLICABLE. 3. APPLY ICE OR FREEZE PACK, IF AVAILABLE. 4. APPLY BAKING SODA, CALAMINE LOTION, OR MEAT TENDERIZER (IF AVAILABLE) TO BITE SITE TO RELIEVE PAIN AND ITCHING. 5. IF THE SITE OF THE BITE IS ON THE FACE, NECK (POSSIBLE AIRWAY PROBLEMS), OR GENITAL AREA, OR IF LOCAL REACTION SEEMS SEVERE, OR IF THE STING IS BY THE DANGEROUS TYPE OF SCORPION FOUND IN THE SOUTHWEST UNITED STATES DESERT, KEEP THE CASUALTY AS QUIET AS POSSIBLE. SEEK MEDICAL ASSISTANCE.
BEE STING	1. IF THE STINGER IS PRESENT, REMOVE BY SCRAPING WITH A KNIFE OR FINGERNAIL. DO NOT SQUEEZE VENOM SAC ON STINGER; MORE VENOM MAY BE INJECTED. 2. REMOVE ALL JEWELRY FROM AFFECTED PART, IF APPLICABLE.

Table 6-1. First Aid Measures for Bites and Stings

TYPES	FIRST AID MEASURES
	3. WASH THE AREA. 4. APPLY ICE OR FREEZE PACK, IF AVAILABLE. 5. IF ALLERGIC SIGNS OR SYMPTOMS APPEAR, BE PREPARED TO PERFORM BASIC LIFE SUPPORT MEASURES. SEEK IMMEDIATE MEDICAL ASSISTANCE.

CHAPTER 7

FIRST AID IN A NUCLEAR, BIOLOGICAL, AND CHEMICAL ENVIRONMENT

7-1. General

American forces have not been exposed to NBC weapons/agents on the battlefield since World War I. In future conflicts and wars we can expect the use of such agents. Nuclear, biological, and chemical weapons will rapidly degrade unit effectiveness by forcing troops to wear protective clothing and by creating confusion and fear. Through training in protective procedures and first aid, units can maintain their effectiveness on the integrated battlefield.

7-2. First Aid Materials

You may be issued the following materials to protect, decontaminate, and use as first aid for NBC exposure. You must know how to use the items; some items are described in *a* through *d* below. It is equally important that you know when to use them.

a. Nerve Agent Pyridostigmine Pretreatment (*NAPP*). You may be issued a blister pack of pretreatment tablets when your commander directs. The NAPP is a pretreatment; it is not an antidote. It improves the effectiveness of the nerve agent antidote. When ordered to take the pretreatment you must take one tablet every 8 hours, mission permitting. This must be taken prior to exposure to nerve agents, since it may take several hours to develop adequate blood levels.

NOTE

Commanders must follow investigational new drug protocols for use of the NAPP.

b. M291 Skin Decontaminating Kit. The M291 Skin Decontaminating Kit (Figure 7-1) contains six packets of XE-555 decontaminant resin.

WARNING

For external use only. May be slightly irritating to the eyes. Keep decontaminating powder out of eyes. Use water to wash toxic agent out of eyes.

c. Nerve Agent Antidote Kit, MARK I. Each service member is issued three MARK Is for use in first aid for nerve agent poisoning (Figure 7-2 and paragraph 7-6).

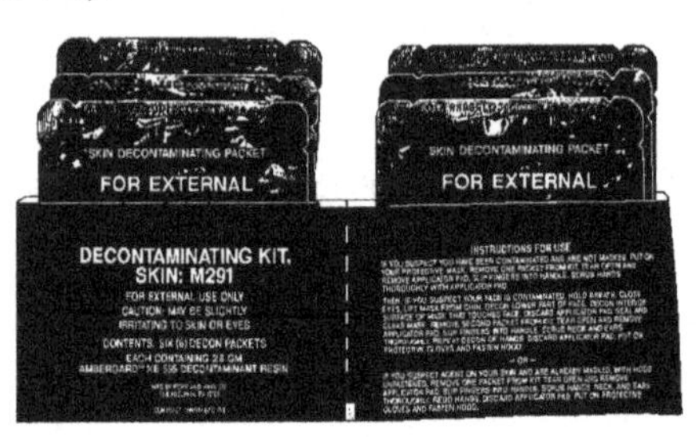

Figure 7-1. M291 Skin Decontamination Kit.

d. Antidote Treatment, Nerve Agent, Autoinjector. A new nerve agent antidote injection device, Antidote Treatment, Nerve Agent, Autoinjector (ATNAA) is currently under development that will replace the MARK I. The ATNAA is a multichambered device with the atropine and pralidoxime chloride in separate chambers. Both antidotes will be administered through a single needle.

7-3. Classification of Chemical and Biological Agents

a. Chemical agents are classified according to the primary physiological effects they produce, such as blistering, choking, vomiting, and incapacitating agents.

b. Biological warfare agents are classified according to the effect they have on man. The effects include their ability to incapacitate and cause death. Most biological warfare agents are delivered as aerosols that effect the respiratory tract; some can be delivered by releasing infected insects, by contaminating food and water, and by injection (injecting material in individuals by terrorist, not mass exposure). These agents are found in living organisms such as fungi, bacteria, and viruses.

WARNING

Swallowing water or food contaminated with nerve, blister, and other chemical agents and with some biological agents can be fatal. NEVER consume water or food that is suspected of being contaminated until it has been tested and found safe for consumption by medical personnel.

7-4. Conditions for Masking Without Order or Alarm

a. Once an attack with a chemical or biological agent is detected or suspected, or information is available that such an agent is about to be used, you must **STOP BREATHING** and mask immediately. **DO NOT WAIT** to receive an order or alarm under the following circumstances:

- Your position is hit by artillery missiles, rockets that produce vapors, smoke, and mists, and aerial sprays.
- Smoke or vapor cloud from an unknown source is present or approaching.
- A suspicious odor, liquid, or solid is present.
- A chemical or biological warfare agent attack is occurring.
- You are entering an area known or suspected of being contaminated.
- When casualties are being received from an area where chemical or biological agents have reportedly been used.
- You have one or more of the following symptoms:
    - An unexplained runny nose.
    - A sudden unexplained headache.
    - A feeling of choking or tightness in the chest or throat.
    - Dimness of vision.
    - Irritation of the eyes.
    - Difficulty in or increased rate of breathing without obvious reasons.
    - Sudden feeling of depression.
    - Dread, anxiety, or restlessness.
    - Dizziness or light-headedness.

- Slurred speech.
- Unexplained laughter or unusual behavior is noted in others.
- Numerous unexplained ill personnel.
- Service members suddenly collapsing without evident cause.
- Animals or birds exhibiting unusual behavior or suddenly dying.

b. For further information on protection and masking procedures, refer to FM 3-4, FM 4-02.7, FM 8-284, and FM 8-285.

7-5. First Aid for a Chemical Attack

Your field protective mask gives protection against biological and chemical warfare agents as well as radiological fallout. With practice you can mask in 9 seconds or less, or put on your mask with hood within 15 seconds.

a. Stop breathing. Don on your mask, seal it properly, and clear and check it; then resume breathing. Give the alarm, and continue the mission. Keep your mask on until the "all clear" signal has been given.

NOTE

Keep your mask on until the area is no longer hazardous and you are told to unmask.

b. If symptoms of nerve agent poisoning (paragraph 7-7) appear, immediately give yourself one MARK I or ATNAA.

CAUTION

Do not inject a nerve agent antidote until you are sure you need it.

c. If your eyes and face become contaminated, you must immediately try to get under cover. You need shelter to prevent further contamination while performing decontamination procedures on your face. If no overhead cover is available, put your poncho over your head before beginning the decontamination process. Then you put on the remaining protective clothing. If vomiting occurs, the mask should be lifted momentarily and drained—with your eyes closed and while holding your breath—then replaced, cleared, and sealed.

d. If nerve agents are used, mission permitting, watch for persons needing nerve agent antidotes and immediately follow procedures outlined in paragraph 7-8*b* or *c*.

e. Decontaminate your skin immediately and clothing and equipment as soon as the mission permits.

7-6. Background Information on Nerve Agents

a. Nerve agents are among the deadliest of chemical agents. Nerve agents enter the body by inhalation, by ingestion, and through the skin. Depending on the route of entry and the amount, nerve agents can produce injury or death within minutes. Nerve agents can achieve their effects with small amounts. Nerve agents are absorbed rapidly, and the effects are felt immediately upon entry into the body. You will be issued three MARK Is or three ATNAAs and one Convulsant Antidote for Nerve Agent (CANA). Each MARK I consists of one atropine autoinjector and one pralidoxime chloride (2 PAM Cl) autoinjector (Figure 7-2A). Each ATNAA consist of a multichambered autoinjector with the atropine and pralidoxime chloride in separate chambers (Figure 7-2C). The CANA is a single autoinjector with flanges (Figure 7-2B). Procedures for use of both the MARK I and ATNAA are described below. You will use either the MARK I or the ATNAA in self-aid and buddy aid as issued.

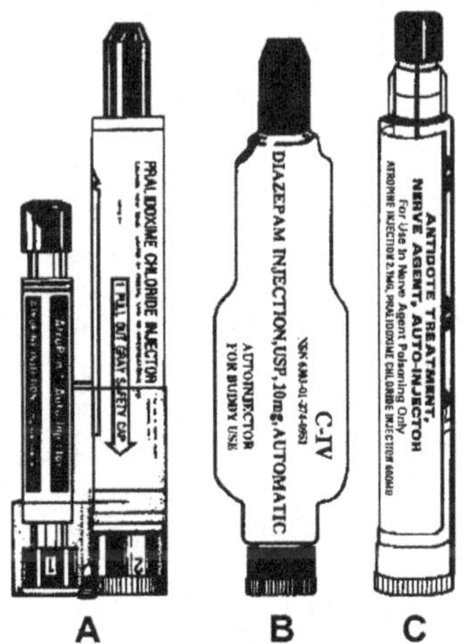

Figure 7-2. Nerve Agent Antidote Kit, MARK I, CANA, and ATNAA.

b. When you have the signs and symptoms of nerve agent poisoning, you should immediately put on the protective mask and then inject yourself with one set of the MARK I or ATNAA. Do not administer the CANA. You should inject yourself in the outer (lateral) thigh muscle (Figure 7-3) or if you are thin, in the upper outer (lateral) part of the buttocks (Figure 7-4).

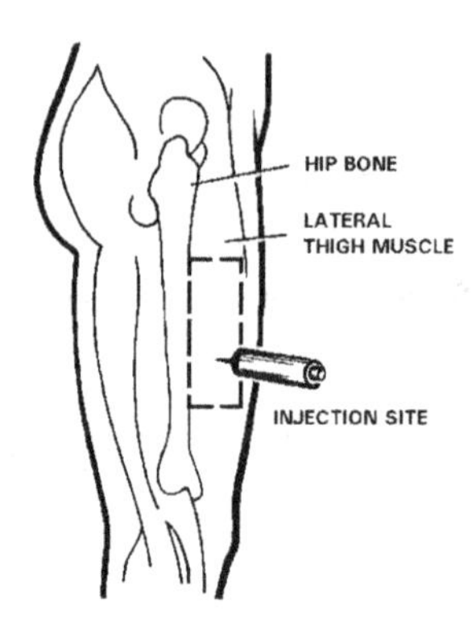

Figure 7-3. Thigh injection site.

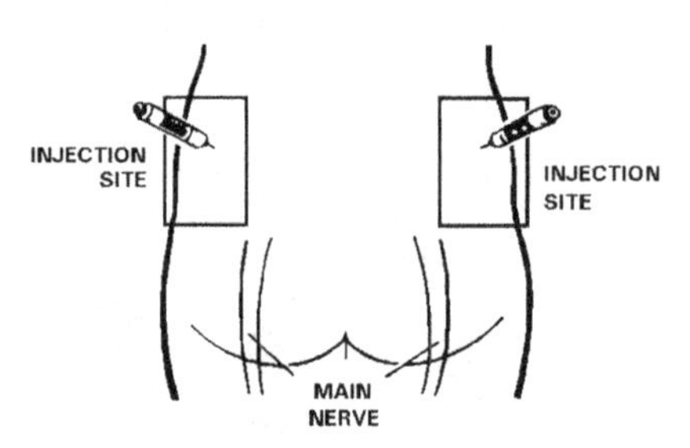

Figure 7-4. Buttocks injection site.

c. Also, you may come upon an unconscious chemical agent casualty who will be unable to care for himself and who will require first aid. You should be able to successfully—

(1) Mask him if he is unmasked.

(2) Inject him, if necessary, with all of **HIS** autoinjectors.

(3) Decontaminate his skin.

(4) Seek medical assistance.

7-7. Signs and Symptoms of Nerve Agent Poisoning

The symptoms of nerve agent poisoning are grouped as **MILD**—those that you recognize and for which you can perform self-aid, and **SEVERE**—those which require buddy aid.

a. MILD Signs and Symptoms.

- Unexplained runny nose.
- Unexplained sudden headache.
- Sudden drooling.
- Difficulty seeing (dimness of vision and miosis).
- Tightness in the chest or difficulty in breathing.
- Localized sweating and muscular twitching in the area of contaminated skin.
- Stomach cramps.
- Nausea.
- Tachycardia followed by bradycardia. (*Tachycardia* is an abnormally rapid heartbeat with a heart rate of over 100 beats per minute. *Bradycardia* is a slow heart rate of less than 60 beats per minute.)

b. SEVERE Signs and Symptoms.

- Strange or confused behavior.
- Wheezing, dyspnea (difficulty in breathing), and coughing.
- Severely pinpointed pupils.
- Red eyes with tearing.
- Vomiting.
- Severe muscular twitching and general weakness.
- Involuntary urination and defecation.

- Convulsions.
- Unconsciousness.
- Respiratory failure.
- Bradycardia.

7-8. First Aid for Nerve Agent Poisoning

First aid for nerve agent poisoning consists of administering the MARK I or ATNAA and CANA.

a. Injection Site. The injection site for administering the antidotes is normally in the outer thigh muscle. The thigh injection site is the area about a hand's width above the knee to a hand's width below the hip joint (Figure 7-3). It is important that the injection be given into a large muscle area. If the individual is thinly built, then the injections should be administered into the upper outer quarter (quadrant) of the buttock (Figure 7-4). Injecting in the buttocks of a thinly built individual avoids injury to the thighbone.

b. Self-Administer MARK I. If you experience any or all of the nerve agent **MILD** symptoms (paragraph 7-7*a*), you must **IMMEDIATELY** put on your protective mask and self-administer one MARK I (Figure 7-2A). Follow the procedure given in Table 7-1. The MARK I is carried in your protective mask carrier, pocket of the MOPP overgarment, or other location as specified in your unit tactical standing operating procedure (TSOP). (In cold weather, the MARK I should be stored in an inside pocket of your clothing to protect the antidote from freezing. A frozen MARK I cannot be immediately used to provide you with antidote, when needed. (However, the MARK I can still be used after complete thawing.)

Table 7-1. Self Aid for Nerve Agent Poisoning

MARK I*	ATNAA*
STEP 1. OBTAIN ONE MARK I.**	STEP 1. OBTAIN ONE ATNAA.**
STEP 2. CHECK INJECTION SITE.	STEP 2. CHECK INJECTION SITE.
STEP 3. HOLD MARK I AT EYE LEVEL WITH NONDOMINANT HAND WITH THE LARGE INJECTOR ON TOP (FIGURE 7-5A).	STEP 3. HOLD ATNAA WITH DOMINANT HAND (FIGURE 7-12A).

Table 7-1. Self Aid for Nerve Agent Poisoning (Continued)

MARK I*	ATNAA*
STEP 4. GRASP SMALL INJECTOR (ATROPINE) (FIGURE 7-5B) AND REMOVE FROM CLIP (FIGURE 7-5C).	STEP 4. GRASP SAFETY CAP WITH NONDOMINANT HAND AND REMOVE FROM INJECTOR (FIGURE 7-12B).
STEP 5. CLEAR HARD OBJECTS FROM INJECTION SITE.	STEP 5. CLEAR HARD OBJECTS FROM INJECTION SITE.
STEP 6. INJECT ATROPINE AT INJECTION SITE APPLYING EVEN PRESSURE TO THE INJECTOR (FIGURE 7-6 OR 7-7). HOLD IN PLACE FOR 10 SECONDS.	STEP 6. INJECT ATNAA AT INJECTION SITE APPLYING EVEN PRESSURE TO THE INJECTOR (FIGURE 7-14 OR 7-15). HOLD IN PLACE FOR 10 SECONDS.
STEP 7. HOLD USED INJECTOR WITH NONDOMINANT HAND.	STEP 7. BEND NEEDLE OF USED INJECTOR BY PRESSING ON A HARD SURFACE TO FORM A HOOK.
STEP 8. GRASP THE LARGE (2 PAM CI) INJECTOR (FIGURE 7-8B) AND PULL IT FROM CLIP (FIGURE 7-8C). DROP CLIP TO GROUND.	STEP 8. ATTACH USED INJECTOR TO BLOUSE POCKET FLAP OF BDO/JSLIST (FIGURE 7-16).
STEP 9. INJECT 2 PAM CI AT INJECTION SITE APPLYING EVEN PRESSURE TO THE INJECTOR (FIGURE 7-6 OR 7-7). HOLD IN PLACE FOR 10 SECONDS.	STEP 9. MASSAGE INJECTION SITE, MISSION PERMITTING.
STEP 10. BEND THE NEEDLES OF ALL USED INJECTORS BY PRESSING ON A HARD SURFACE TO FORM A HOOK.	
STEP 11. ATTACH ALL USED INJECTORS TO BLOUSE POCKET FLAP OF BDO/JSLIST (FIGURE 7-9).	
STEP 12. MASSAGE INJECTION SITE, MISSION PERMITTING.	

* USE STEPS LISTED FOR TYPE OF ANTIDOTE DEVICE ISSUED.

** ONLY ADMINISTER ONE MARK I OR ATNAA AS SELF-AID. DO NOT SELF-ADMINISTER CANA.

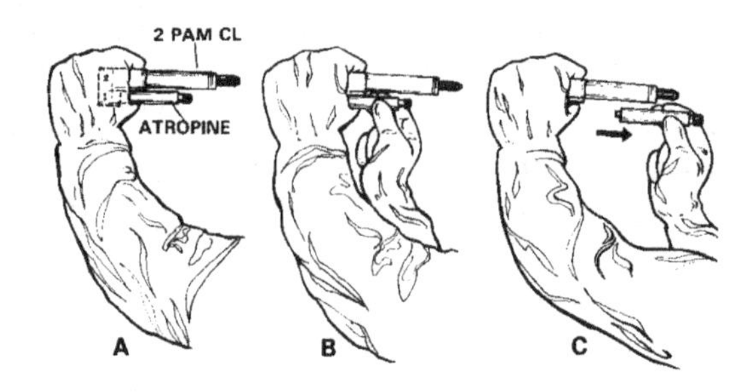

Figure 7-5. Removing the atropine autoinjector from the MARK I clip.

CAUTION

DO NOT cover or hold the needle end with your hand, thumb, or fingers—you might accidentally inject yourself. An accidental injection into the hand **WILL NOT** deliver an effective dose of the antidote, especially if the needle goes through the hand.

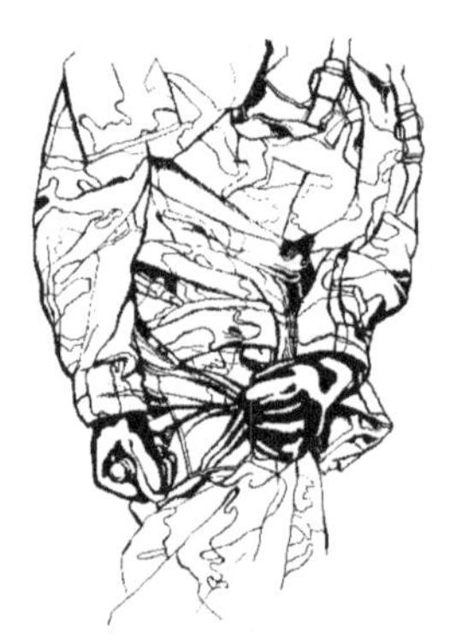

Figure 7-6. Thigh injection site for self-aid.

NOTE

If you are thinly built, inject yourself into the upper outer quadrant of the buttock (Figure 7-7). There is a nerve that crosses the buttocks; hitting this nerve can cause paralysis. Therefore, you must only inject into the *upper outer quadrant* of the buttock.

Figure 7-7. Buttocks injection site for self-aid.

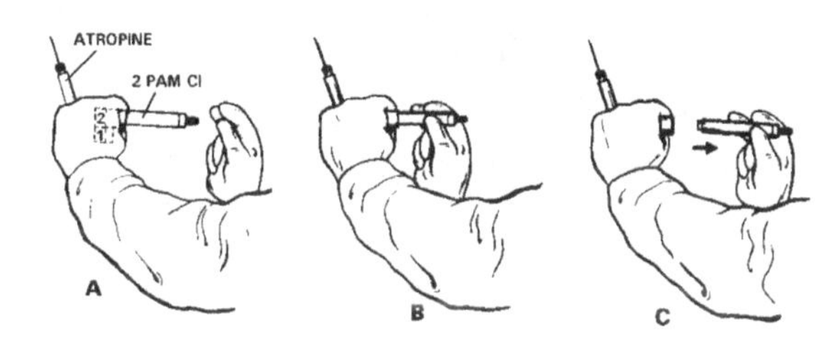

Figure 7-8. Removing the 2 PAM Cl autoinjector from the MARK I clip.

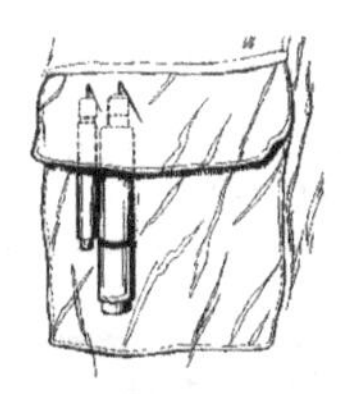

Figure 7-9. One set of used MARK I autoinjectors attached to pocket flap.

NOTES

1. **DO NOT** give yourself another set of injections. If you are able to walk without assistance, know who you are, and where you are, you **WILL NOT** need the second set of injections. (If not needed, giving yourself a second set of MARK I injections or ATNAA may create a nerve agent antidote overdose, which could cause incapacitation [inability to perform mission or defend yourself].)

2. If you continue to have symptoms of nerve agent poisoning, seek someone else (a buddy) to check your symptoms and administer the additional sets of injections, if required.

c. Buddy Evaluation and Buddy Aid. Service members may seek assistance after self-aid (self-administering one MARK I or ATNAA) or may become incapacitated after self-aid. A buddy must evaluate the individual to determine if additional antidotes are required to counter the effects of the nerve agent. Also, service members may experience **SEVERE** symptoms of nerve agent poisoning (paragraph 7-7*b*); they will not be able to treat themselves. In either case, other service members must perform buddy aid as quickly as possible. Before initiating buddy aid, determine if one set of MARK I autoinjectors has already been used so that no more than three sets of the antidote are administered. Buddy aid also includes administering the CANA with the third MARK I or ATNAA to prevent convulsions. Follow the procedures indicated in Table 7-2.

WARNING

Squat, DO NOT kneel, when masking the casualty or administering the nerve agent antidote to the casualty. Kneeling may force the chemical agent into or through your protective clothing.

CAUTION

DO NOT use your own MARK I, ATNAA, or CANA on a casualty. If you use your own, you may not have any antidote if needed for self-aid.

WARNING

DO NOT inject into areas close to the hip, knee, or thigh-bone.

Table 7-2. Buddy Aid/Combat Lifesaver Aid for Nerve Agent Casualty.

MARK I*	ATNAA*	CANA**
STEP 1. MASK THE CASUALTY AND POSITION HIM ON HIS SIDE (SWIMMER'S POSITION).	STEP 1. MASK THE CASUALTY AND POSITION HIM ON HIS SIDE (SWIMMER'S POSITION).	STEP 1. OBTAIN BUDDY'S CANA.
STEP 2. POSITION YOURSELF NEAR THE CASUALTY'S THIGH.	STEP 2. POSITION YOURSELF NEAR THE CASUALTY'S THIGH.	STEP 2. CHECK INJECTION SITE.

Table 7-2. Buddy Aid/Combat Lifesaver Aid for Nerve Agent Casualty (Continued).

MARK I*	ATNAA*	CANA**
STEP 3. OBTAIN BUDDY'S THREE OR REMAINING MARK Is.	STEP 3. OBTAIN BUDDY'S THREE OR REMAINING ATNAAs.	STEP 3. HOLD CANA IN A CLOSED FIST WITH DOMINANT HAND (FIGURE 7-12A).
STEP 4. CHECK INJECTION SITE.	STEP 4. CHECK INJECTION SITE.	STEP 4. GRASP SAFETY CAP WITH NONDOMINANT HAND AND REMOVE FROM INJECTOR (FIGURE 7-12B).
STEP 5. HOLD MARK I WITH NONDOMINANT HAND (FIGURE 7-5A).	STEP 5. HOLD ATNAA IN A CLOSED FIST WITH DOMINANT HAND (FIGURE 7-12A).	STEP 5. CLEAR HARD OBJECTS FROM INJECTION SITE.
STEP 6. GRASP SMALL INJECTOR (ATROPINE) AND REMOVE FROM CLIP (FIGURE 7-5B).	STEP 6. GRASP SAFETY CAP WITH NONDOMINANT HAND AND REMOVE FROM INJECTOR (FIGURE 7-12B).	STEP 6. INJECT CANA AT INJECTION SITE BY APPLYING EVEN PRESSURE TO THE INJECTOR, NOT A JABBING MOTION (FIGURE 7-14 OR 7-15). HOLD IN PLACE FOR 10 SECONDS.
STEP 7. CLEAR HARD OBJECTS FROM INJECTION SITE.	STEP 7. CLEAR HARD OBJECTS FROM INJECTION SITE.	STEP 7. BEND NEEDLE OF INJECTOR BY PRESSING ON A HARD SURFACE TO FORM A HOOK.
STEP 8. INJECT ATROPINE AT INJECTION SITE BY APPLYING EVEN PRESSURE TO THE INJECTOR, NOT A JABBING MOTION (FIGURE 7-10 OR 7-11). HOLD IN PLACE FOR 10 SECONDS.	STEP 8. INJECT ATNAA AT INJECTION SITE BY APPLYING EVEN PRESSURE TO THE INJECTOR, NOT A JABBING MOTION (FIGURE 7-14 OR 7-15). HOLD IN PLACE FOR 10 SECONDS.	STEP 8. ATTACH USED INJECTOR TO BLOUSE POCKET FLAP OF BDO/JSLIST (FIGURE 7-16).
STEP 9. HOLD USED INJECTOR BETWEEN LITTLE FINGER AND RING FINGER OF NONDOMINANT HAND (FIGURE 7-5A).	STEP 9. BEND NEEDLE OF INJECTOR BY PRESSING ON A HARD SURFACE TO FORM A HOOK.	STEP 9. MASSAGE INJECTION SITE, MISSION PERMITTING.
STEP 10. PULL LARGE INJECTOR (2 PAM CI) FROM CLIP (FIGURE 7-5C). DROP CLIP TO GROUND.	STEP 10. ATTACH ALL USED INJECTORS TO BLOUSE POCKET FLAP OF BDO/JSLIST (FIGURE 7-16).	

Table 7-2. Buddy Aid/Combat Lifesaver Aid for Nerve Agent Casualty (Continued).

MARK I*	ATNAA*	CANA**
STEP 11. INJECT 2 PAM CI AT INJECTION SITE BY APPLYING EVEN PRESSURE TO THE INJECTOR, NOT A JABBING MOTION (FIGURE 7-10 OR 7-11). HOLD IN PLACE FOR 10 SECONDS.	STEP 11. MASSAGE INJECTION SITE, MISSION PERMITTING.	
STEP 12. REPEAT STEPS ABOVE FOR REMAINING MARK Is.		
STEP 13. BEND THE NEEDLES OF ALL USED INJECTORS BY PRESSING ON A HARD SURFACE TO FORM A HOOK.		
STEP 14. ATTACH ALL USED INJECTORS TO BLOUSE POCKET FLAP OF BDO/JSLIST (FIGURE 7-13).		
STEP 15. MASSAGE INJECTION SITE, MISSION PERMITTING.		

* USE STEPS LISTED FOR TYPE OF ANTIDOTE DEVICE ISSUED.

** CANA IS USED IN BUDDY AID/CLS AID ONLY. DO NOT USE IN SELF-AID.

NOTE

If the casualty is thinly built, inject the antidote into the buttock. Only inject the antidote into the upper outer portion of the casualty's buttock (Figure 7-11). This avoids hitting the nerve that crosses the buttocks (Figure 7-4). Hitting this nerve can cause paralysis.

Figure 7-10. Injecting the casualty's thigh (Mark I or CANA).

Figure 7-11. Injecting the casualty's buttocks (Mark I or CANA).

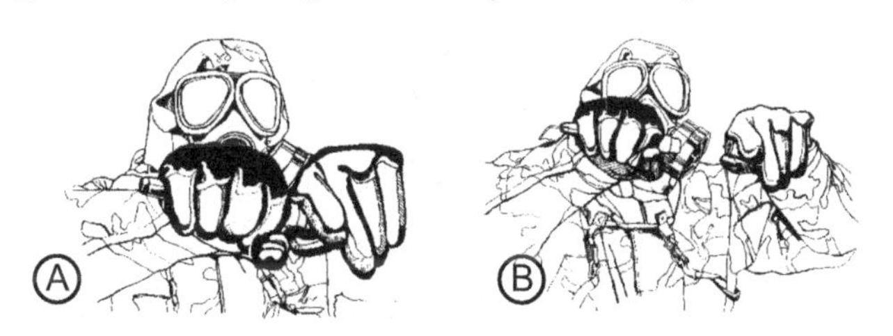

Figure 7-12. Preparing CANA or ATNAA for injection.

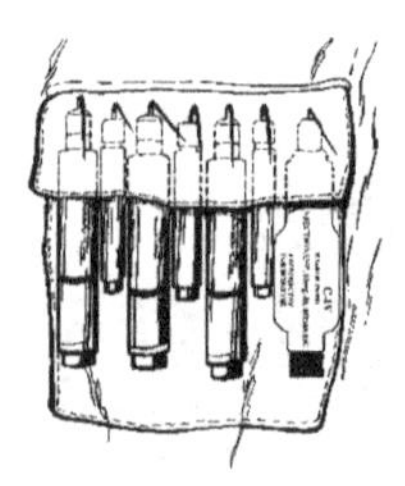

Figure 7-13. Three sets of used MARK I autoinjectors and one CANA autoinjector attached to pocket flap.

d. Self-Administer Antidote Treatment Nerve Agent Autoinjector. If you experience any or all of the nerve agent **MILD** symptoms (paragraph 7-7*b*), you must **IMMEDIATELY** self-administer one ATNAA following the procedure given Table 7-1.

NOTE

If you are thinly-built, inject yourself into the upper outer quarter (quadrant) of the buttock (Figure 7-15). There is a nerve that crosses the buttocks; hitting this nerve can cause paralysis. Therefore, you must only inject into the upper outer quarter (quadrant) of the buttocks.

Figure 7-14. Self-administration of ATNAA (thigh).

Figure 7-15. Self-administration of ATNAA (buttock).

NOTE

If you continue to have symptoms of nerve agent poisoning, seek someone else (a buddy) to check your symptoms and administer your remaining sets of injections, if required.

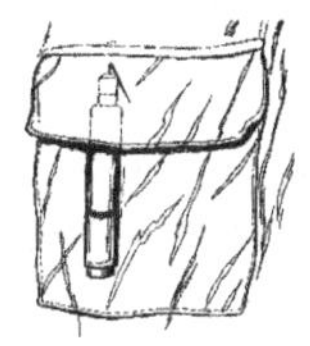

Figure 7-16. Used ATNAA attached to clothing.

e. Buddy Assistance. Service members may seek assistance after self-aid (self-administering one ATNAA) or may become incapacitated after self-aid. A buddy must evaluate the individual to determine if additional antidotes are required to counter the effects of the nerve agent. Also, service members may experience **SEVERE** symptoms of nerve agent poisoning (paragraph 7-7*b*); they will not be able to treat themselves. In either case, other service members must perform buddy aid as quickly as possible. Before initiating buddy aid, determine if one ATNAA has already been used so that no more than three ATNAA are administered. Buddy aid also includes administering the CANA with the third ATNAA to prevent convulsions. Follow the procedures indicated in Table 7-2.

WARNING

Squat, DO NOT kneel, when masking the casualty or administering the nerve agent antidotes to the casualty. Kneeling may force any chemical agent on your overgarment into or through your protective clothing.

Figure 7-17. Buddy injecting casualty's outer thigh (ATNAA or CANA).

NOTE

If the casualty is thinly built, inject the antidote into the buttocks (Figure 7-18). Only inject the antidote into the upper outer portion of the casualty's buttocks. This avoids hitting the nerve that crosses the buttocks (Figure 7-4). Hitting this nerve can cause paralysis.

WARNING

DO NOT inject into areas close to the hip, knee, or thighbone.

Figure 7-18. Buddy injecting casualty's buttocks (ATNAA or CANA).

Figure 7-19. Three used ATNAAs and one CANA autoinjector attached to clothing.

f. Combat Lifesaver.

(1) The combat lifesaver must check to verify if the individual has received three sets of MARK I or ATNAAs. If not, the combat lifesaver performs first aid as described for buddy aid above. If the

individual has received the initial three sets of MARK I, then the combat lifesaver may administer additional atropine injections at approximately 15 minute intervals until atropinization is achieved (that is a heart rate above 90 beats per minute, reduced bronchial secretions, and reduced salivations). Administer additional atropine at intervals of 30 minutes to 4 hours to maintain atropinization or until the casualty is placed under the care of medical personnel. Check the heart rate by lifting the casualty's mask hood and feeling for a pulse at the carotid artery. Request medical assistance as soon as the tactical situation permits.

(2) The CLS should administer additional CANA to casualties suffering convulsions. Administer a second, and if needed, a third CANA at 5 to 10 minute intervals for a maximum of three injections (30 milligrams diazepam). Follow the steps and procedures described in buddy aid for administering the CANA. **DO NOT** give more than two additional injections for a total of three (one self-aid plus two by the CLS).

7-9. Blister Agents

Blister agents (vesicants) include mustard (H and HD), nitrogen mustards (HN), lewisite (L), and other arsenicals, mixtures of mustards and arsenicals, and phosgene oxime (CX). Blister agents may act on the eyes, mucous membranes, lungs, and skin. They burn and blister the skin or any other body parts they contact. Even relatively low doses may cause serious injury. Blister agents damage the respiratory tract (nose, sinuses, and windpipe) when inhaled and cause vomiting and diarrhea when absorbed. Lewisite and CX cause immediate pain on contact. However, mustard agents are deceptive as there is little or no pain at the time of exposure. Thus, in some cases, signs of injury may not appear for several hours after exposure.

a. Protective Measures. Your protective mask with hood and protective overgarment provide protection against blister agents. If it is known or suspected that blister agents are being used, **STOP BREATHING**, put on your mask and your protective overgarment.

CAUTION

Large drops of liquid vesicants on the protective overgarment ensemble may penetrate it if allowed to stand for an extended period. Remove large drops as soon as possible.

b. Signs and Symptoms of Blister Agent Poisoning.

(1) Immediate and intense pain upon contact with L, LH (lewisite and mustard) mixture, and CX. No initial pain upon contact with mustard.

(2) Inflammation and blisters (burns) resulting in tissue destruction. The severity of a chemical burn is directly related to the concentration of the agent and the duration of contact with the skin. The longer the agent is in contact with the tissue, the more serious the injury will be.

(3) Vomiting and diarrhea. Exposure to high concentrations of vesicants may cause vomiting or diarrhea.

(4) Death. The blister agent vapors absorbed during ordinary field exposure will probably not cause enough internal body (systemic) damage to result in death. However, death may occur from prolonged exposure to high concentrations of vapor or from extensive liquid contamination over wide areas of the skin, particularly *when decontamination is neglected or delayed.*

c. First Aid Measures.

(1) Use your M291 Skin Decontaminating Kit to decontaminate your skin and use water to flush contaminated eyes. Decontamination of vesicants must be done immediately (within 1 minute is best).

(2) If blisters form, cover them loosely with a field dressing and secure the dressing.

CAUTION

Blisters are actually burns. **DO NOT** attempt to decontaminate the skin where blisters have formed, as the agent has already been absorbed.

(3) If you receive blisters over a wide area of the body, you are considered seriously burned. Seek medical assistance immediately.

(4) If vomiting occurs, the mask should be lifted momentarily and drained—while the eyes are closed and the breath is held—and replaced, cleared, and sealed.

(5) Remember, if vomiting or diarrhea occurs after having been exposed to blister agents, seek medical assistance immediately.

7-10. Choking Agents (Lung-Damaging Agents)

Chemical agents that attack lung tissue, primarily causing fluid buildup (pulmonary edema), are classified as choking agents (lung-damaging agents). This group includes phosgene (CG), diphosgene (DP), chlorine (Cl), and chloropicrin (PS). Of these four agents, CG is the most dangerous and is more likely to be employed by the enemy in future conflict.

a. Protective Measures. Your protective mask gives adequate protection against choking agents.

b. Signs and Symptoms. During and immediately after exposure to choking agents (depending on agent concentration and length of exposure), you may experience some or all of the following signs and symptoms:

- Tears (lacrimation).
- Coughing.
- Choking.
- Tightness of chest.
- Nausea and vomiting.
- Headaches.

c. Self-Aid.

(1) The protective mask should be put on immediately when any of the conditions described in *b* above exist. Another indication of a CG attack is an odor like newly mown hay; however, **DO NOT** rely upon odor as indication of a chemical attack.

(2) If some CG is inhaled, normal combat duties should be continued unless there is difficulty in breathing, nausea, vomiting, or more than the usual shortness of breath during exertion. If any of the above symptoms occur and the mission permits, remain at quiet rest until medical evacuation is accomplished.

d. Death. With ordinary field exposure to choking agents, death will probably not occur. *However, prolonged exposure to high concentrations of the vapor and neglect or delay in masking can be fatal.*

7-11. Cyanogen (Blood) Agents

Cyanogen agents interfere with proper oxygen utilization in the body. Hydrogen cyanide (AC) and cyanogen chloride (CK) are the primary agents in this group.

a. Protective Measures. Your protective mask with a fresh filter gives adequate protection against field concentrations of cyanogen agent vapor. The protective overgarments, as well as the mask, are needed when exposed to liquid AC.

b. Signs and Symptoms. During and immediately after exposure to cyanogen agents (depending on agent concentration and length of exposure), you may experience some or all of the following signs and symptoms:

- Tearing (lacrimation).
- Eye, nose, and throat irritation.
- Sudden stimulation of breathing (unable to hold breath).
- Nausea.
- Coughing.
- Tightness of chest.
- Headache.
- Light-headedness (dizziness).
- Unconsciousness.

c. First Aid.

(1) *Hydrogen cyanide.* During any chemical attack, if you get a sudden stimulation of breath or detect an odor like bitter almonds, **PUT ON YOUR MASK IMMEDIATELY**. Speed is absolutely essential since this agent acts so rapidly that within a few seconds its effects will make it impossible for service members to put on their mask by themselves. Stop breathing until the mask is on, if at all possible. This may be very difficult since the agent strongly stimulates respiration.

(2) *Cyanogen chloride.* Put your mask on immediately if you experience any irritation of the eyes, nose, or throat. Service members

who are unable to mask should be masked by the nearest service member (buddy).

d. *Medical Assistance*. If you suspect that you have been exposed to blood agents, seek medical assistance immediately.

7-12. Incapacitating Agents

An incapacitating agent is a chemical agent which produces temporary, disabling conditions which persist for hours to days after exposure. Unlike riot control agents, which usually are momentary or fleeting in action, incapacitating agents have a persistent effect. It is likely that smoke-producing munitions or aerosols will disseminate such agents, thus making breathing their means of entry into the body. The protective mask is, therefore, essential.

a. There are no specific first aid measures to relieve the symptoms of incapacitating agents. Supportive first aid and physical restraint may be indicated. If the casualty is stuporous or comatose, be sure that respiration is unobstructed; then turn him on his side in case vomiting should occur. Complete cleansing of the skin with soap and water should be done as soon as possible; or, the M291 Skin Decontaminating Kit can be used if washing is impossible. Remove weapons and other potentially harmful items from service members who are suspected of having these symptoms. Harmful items include cigarettes, matches, medications, and small items that might be swallowed accidentally. Delirious (confused) persons have been known to attempt to eat items bearing only a superficial resemblance to food.

b. Incapacitating agents (anticholinergic drugs BZ type) may produce alarming dryness and coating of the lips and tongue; however, there is usually no danger of immediate dehydration. Fluids should be given sparingly, if at all, because of the danger of vomiting and because of the likelihood of temporary urinary retention due to paralysis of bladder muscles.

c. If the body temperature is elevated and mucous membranes are dry, immediate and vigorous cooling (as for heatstroke) is indicated. Methods that can be used to cool the skin are spraying with cool water and air circulation (fanning); applying alcohol soaked cloths and air circulation; and providing maximum exposure to air in a shaded area, along with maximum air circulation. Such cases are usually a result of anticholinergic poisoning. Rapid evacuation should be accomplished since medical treatment with the appropriate medication may be lifesaving.

CAUTION

DO NOT use **ice** for cooling the skin.

d. Reassurance and a firm, but friendly, attitude by individuals providing first aid will be beneficial if the casualty appears to comprehend what is being said. Conversation is a waste of time if the service member is incoherent or cannot understand what is being said. In such cases, the less said, the better it is—these casualties will benefit more from prompt and vigorous restraint and evacuation to an MTF.

7-13. Incendiaries

Incendiaries can be grouped as WP, thickened gasoline, metal, and oil and metal. You must learn to protect yourself against these incendiaries.

a. White phosphorus is used primarily as a smoke producer but can be used for its incendiary effect to ignite field expedients and combustible materials. The burns from WP are usually multiple, deep, and variable in size. When particles of WP get on the skin or clothing, they continue to burn until deprived of air. They also have a tendency to stick to a surface and must be brushed off or picked out.

(1) If burning particles of WP strike and stick to your clothing, quickly take off the contaminated clothing before the WP burns through to the skin.

(2) If burning WP strikes your skin, smother the flame with water, a wet cloth, or mud.

NOTE

Since WP is soluble in oil, **DO NOT** use grease, oily ointments, or eye ointments to smother the flame.

(3) Keep the WP particles covered with a wet material to exclude air until you can remove them or have them removed from your skin.

(4) Remove the WP particles from the skin by brushing them with a wet cloth and by picking them out with a knife, bayonet, stick, or other available object.

(5) Seek medical assistance when the mission permits.

b. Thickened fuel mixtures (napalm) have a tendency to cling to clothing and body surfaces, thereby producing prolonged exposure and severe burns. The first aid for these burns is the same as for other heat burns. The heat and irritating gases given off by these combustible mixtures may cause lung damage, which must be treated by medical personnel.

c. Metal incendiaries pose special problems. Thermite particles on the skin should be immediately cooled with water and then removed. The first aid for these burns is the same as for other heat burns. Particles of magnesium on the skin burn quickly and deeply. Like other metal incendiaries, they must be removed. Ordinarily, medical personnel should do the complete removal of these particles as soon as possible. Immediate medical treatment is required.

d. Oil and metal incendiaries have much the same effect on contact with the skin and clothing as those discussed (*b* and *c* above). First aid measures for burns are discussed in Chapter 3.

7-14. Biological Agents and First Aid

a. Biological attacks can result in combat ineffectiveness by introducing disease-causing organisms into a troop population.

b. Once a disease is identified, first aid or medical treatment is initiated, depending on the seriousness of the disease. First aid measures are concerned with observable symptoms of the disease such as diarrhea or vomiting.

7-15. Toxins

Toxins are alleged to have been used in past conflicts. Witnesses and victims have described the agent as toxic rain (or yellow rain) because it was reported to have been released from aircraft as a yellow powder or liquid that covered ground, structures, vegetation, and people.

a. *Signs and Symptoms.* The occurrence of the symptoms from toxins may appear in a period of a few minutes to several hours depending on the particular toxin, the service member's susceptibility, and the amount of toxin inhaled, ingested, or deposited on the skin. Symptoms from toxins usually involve the central nervous system but are often preceded by less prominent symptoms, such as nausea, vomiting, diarrhea, cramps, or stomach irritation and burning sensation. Typical neurological symptoms often develop rapidly in severe cases; for example, visual disturbances, inability to swallow, speech difficulty, lack of muscle coordination, and sensory abnormalities (numbness of mouth, throat, or extremities). Yellow rain (mycotoxins) also may have hemorrhagic symptoms, which could include any or all of the following:

- Dizziness.
- Severe itching or tingling of the skin.
- Formation of multiple, small, hard blisters.
- Coughing up blood.
- Shock (which could result in death).

b. *Self-Aid.* Upon recognition of an attack employing toxins, you must immediately take the following actions:

(1) Stop breathing, put on your protective mask with hood, and then resume breathing. Next, put on your protective clothing.

(2) Should severe itching of the face become unbearable, quickly—

- Loosen the cap on your canteen.
- Take and hold a deep breath and lift your mask.
- While holding your breath, close your eyes and flush your face with generous amounts of water.

CAUTION

DO NOT rub or scratch your eyes. Try not to let the water run onto your clothing or protective overgarment.

- Put your protective mask back on, seat it properly, clear it, and check it for a seal; then resume breathing.
- Decontaminate your skin by bathing with soap and water as soon as the mission permits.

- Change clothing and decontaminate your protective mask using soap and water. Replace the filters if directed.

(3) If vomiting occurs, the mask should be lifted momentarily and drained—while the eyes are closed and the breath is held—and replaced, cleared, and sealed.

c. *Medical Assistance.* If you suspect that you have been exposed to toxins, you should seek medical assistance immediately.

7-16. Nuclear Detonation

a. Three types of injuries may result from a nuclear detonation. These are thermal, blast, and radiation injuries. Many times the casualty will have a combination of these types of injuries. First aid for thermal and blast injuries is provided based on observable injuries, such as burns, hemorrhage, or fractures.

b. The signs and symptoms of radiation illness in the initial phase include the rapid onset of nausea, vomiting, and malaise (tiredness). The only first aid procedure for radiological casualties is decontamination.

CHAPTER 8

FIRST AID FOR PSYCHOLOGICAL REACTIONS

8-1. General

Psychological first aid is as natural and reasonable as physical first aid and is just as familiar. When you were hurt as a child, the understanding attitude of your parents did as much as the psychological effect of a bandage. Later, your disappointment or grief was eased by supportive words from a friend. Certainly, taking a walk and talking things out with a friend are familiar ways of dealing with an emotional crisis. The same natural feelings that make us want to help a person who is injured make us want to give a helping hand to a buddy who is upset. *Psychological first aid* really means nothing more complicated than assisting people with emotional distress whether it results from physical injury, disease, or excessive stress. Emotional distress is not always as visible as a wound or a broken bone. However, overexcitement, severe fear, excessive worry, deep depression, misdirected irritability, and anger are signs that stress has reached the point of interfering with effective coping. The more noticeable the symptoms become, the more urgent the need for you to be of help and the more important it is for you to know *how* to help.

8-2. Importance of Psychological First Aid

You must know how to give psychological first aid to be able to help yourself, your buddies, and your unit in order to keep performing the mission. Psychological first aid measures are simple and easy to understand. Your decision of what to do depends upon your ability to observe the service member and understand his needs. Making the best use of resources requires ingenuity on your part. A stress reaction resulting in poor judgment can cause injury or even death to yourself or others on the battlefield. It can be even more dangerous if other persons are affected by the judgment of an emotionally upset service member. If it is detected early enough, the affected service member stands a good chance of remaining in his unit as an effective member. If it is not detected early and if the service member becomes more emotionally upset, he may become a threat to himself and to others.

8-3. Situations Requiring Psychological First Aid

- Psychological first aid (buddy aid) is most needed at the first sign that a service member cannot perform the mission because of emotional distress. Stress is inevitable in combat, in hostage and terrorist situations, and in civilian disasters such as floods, hurricanes, or industrial accidents. Most emotional reactions to such situations are temporary, and the service member can still carry on with encouragement. Painful or disruptive symptoms may last for minutes, hours, or days. However, if the stress symptoms are seriously disabling, they may be psychologically contagious and endanger not only the emotionally upset service member but also the entire unit.

- Sometimes people continue to function well during a disastrous event, but suffer from emotional scars which impair their job performance or quality of life at a later time. Painful memories and dreams may recur for months and years and still be considered a normal reaction. However, if the memories are so painful that the person must avoid all situations which arouse them, becomes socially withdrawn, or shows symptoms of anxiety, depression, or substance abuse, he needs treatment. Experience with police, firemen, emergency medical technicians, and others who deal with disasters has proved that the routine application of psychological first aid to all the participants, including those who have functioned well, greatly reduces the likelihood of future serious post-traumatic stress disorders (PTSDs).

8-4. Interrelationship of Psychological and Physical First Aid

Psychological first aid should go hand in hand with physical first aid. The discovery of a physical injury or cause for an inability to function does not rule out the possibility of a psychological injury (or vice versa). The person suffering from pain, shock, fear of serious injury, or fear of death does not respond well to joking, indifference, or fearful-tearful attention. Fear and anxiety may take as high a toll of the service member's strength as does the loss of blood.

8-5. Goals of Psychological First Aid

The goals of psychological first aid are to—

- Be supportive; assist the service member in dealing with his stress reaction.

- Prevent, and if necessary control, behavior harmful to himself and to others.

- Return the service member to duty as soon as possible after dealing with the stress reaction.

8-6. Respect for Others' Feelings

a. Accept the service member you are trying to help without censorship or ridicule. Respect his right to his own feelings. Even though your feelings, beliefs, and behavior are different, DO NOT blame or make light of him for the way he feels or acts. Your purpose is to help him in this tough situation, not to be his critic. A person DOES NOT WANT to be upset and worried. When he seeks help, he needs and expects consideration of his fears, not abrupt dismissal or ridicule.

b. Realize that people are the products of a wide variety of factors. All people DO NOT react the same way to the same situations. Each individual has complex needs and motivations, both conscious and unconscious, that are uniquely his own. Often the one thing that finally causes the person to become overloaded by a stressful situation is not the stressor itself, but some other problem.

8-7. Emotional and Physical Disability

a. Accept emotional disability as being just as real as physical disability. If a service member's ankle is seriously sprained in a fall, no one expects him to run right away. A service member's emotions may be temporarily strained by the overwhelming stress of battle or other traumatic incident. DO NOT demand that he pull himself together immediately and carry on without a break. Some individuals can pull themselves together immediately, but others cannot. The service member whose emotional stability has been disrupted has a disability just as real as the service member who has sprained his ankle. There is an unfortunate tendency in many people to regard as real only what they can see, such as a wound or bleeding. Some people tend to assume that damage involving a person's mind and emotions is just imagined, that he is not really sick or injured, and that he could overcome his trouble by using his will power.

b. The terms *it's all in your head, snap out of it,* and *get control of yourself* are often used by people who believe they are being helpful. Actually, these terms are expressions of hostility because they show lack of understanding. They only emphasize weakness and inadequacy. Such terms are of no use in psychological first aid.

c. Every physically injured person has some emotional reaction to the fact that he is injured.

(1) It is normal for an injured person to feel upset. The more severe the injury, the more insecure and fearful he becomes, especially

if the injury is to a body part which is highly valued. For example, an injury to the eyes or the genitals, even though relatively minor, is likely to be extremely upsetting. An injury to some other part of the body may be especially disturbing to an individual for his own particular reason. For example, an injury of the hand may be a terrifying blow to a surgeon or an injury to the eye of a pilot.

(2) An injured service member always feels less secure, more anxious, and more afraid not only because of what has happened to him but because of what he imagines may happen as a result of his injury. This fear and insecurity may cause him to be irritable, uncooperative, or unreasonable. As you help him, always keep in mind that such behavior has little or nothing to do with you personally. He needs your patience, reassurance, encouragement, and support.

8-8. Combat and Other Operational Stress Reactions

Stress reaction is a temporary emotional disorder or inability to function, experienced by a previously normal service member as a reaction to the overwhelming or cumulative stress of combat. Stress reaction gets better with reassurance, rest, physical replenishment, and activities that restore confidence. All service members are likely to feel stress reaction under conditions of intense and/or prolonged stress. They may even become stress reaction casualties, unable to perform their mission for hours or days. Other combat and operational stress reactions (COSRs) may result in negative behavior, but are not termed *stress reaction,* as they need more intensive treatment. These negative COSRs may result in misconduct stress behaviors such as drug and alcohol abuse, criminal acts, looting, desertion, and self-inflicted wounds. These harmful COSRs can often be prevented by good psychological first aid. Service members who commit misconduct stress behaviors may require disciplinary action rather than medical treatment.

8-9. Reactions to Stress

Most service members react to stressful incidents after the situation has passed. All service members feel some fear. This fear may be greater than they have experienced at any other time, or they may be more aware of their fear. In such a situation, they should not be surprised if they feel shaky or become sweaty, nauseated, or confused. These reactions are normal and are not a cause for concern. However, some reactions, either short- or long-term, will cause problems if left unchecked. See paragraph 8-13 for more information.

a. Emotional Reactions.

(1) The most obvious combat stress reaction (CSR) is inefficient performance. This can be demonstrated by—

- Slow thinking (or reaction time).
- Difficulty recognizing priorities and seeing what needs to be done.
- Difficulty getting started.
- Indecisiveness and having trouble focusing attention.
- Tendency to do familiar tasks and be preoccupied with familiar details. (This can reach the point where the person is very passive, such as just sitting or wandering about not knowing what to do.)

(2) A less common reaction may be uncontrolled emotional outbursts; this can be demonstrated by crying, screaming, or laughing. Some service members will react in the opposite way. They will be very withdrawn and silent and try to isolate themselves from everyone. These service members should be encouraged to remain with their assigned unit. Uncontrolled reactions may appear by themselves or in any combination (the person may be crying uncontrollably one minute and then laughing the next). In this state, the person is restless and cannot keep still. He may run about, apparently without purpose. Inside, he feels a great rage or fear and his physical acts may show this. In his anger he may indiscriminately strike out at others.

b. *Loss of Adaptability.*

(1) In a desperate attempt to get away from the danger, which has overwhelmed him, a service member may panic and become confused. His mental ability may be so impaired he cannot think clearly or even follow simple commands. His judgment may be faulty and he may not be aware of his actions, such as standing up in his fighting position during an attack.

(2) In other cases, overwhelming stress may produce symptoms that are often associated with head injuries. For example, the service member may appear dazed or be found wandering around aimlessly. He may appear confused and disoriented and may seem to have a complete or partial loss of memory. In such cases, especially when no eyewitnesses can provide evidence that the service member has NOT suffered a head injury, it is necessary for him to be rapidly medically evacuated. **DO NOT** allow the

service member to expose himself to further personal danger until the cause of the problem has been determined.

c. Sleep Disturbance and Repetition of Dreams. A person who has been overwhelmed by stress often has difficulty sleeping. The service member may experience nightmares related to the stressors. Remember that nightmares, in themselves, are not considered abnormal when they occur soon after a period of intensive stress. As time passes, the nightmares usually become less frequent and less intense. In extreme cases, a service member, even when awake, may think repeatedly of the incident, feel as though it is happening again, and act out parts of his stress over and over again. For some persons, this repetitious reexperiencing of the stressful event may be necessary for eventual recovery; therefore, it should not be discouraged or viewed as abnormal. For the person reexperiencing the event, such reaction may be disruptive. The service member needs to be encouraged to *ventilate* about the incident. Ventilation is a technique where the service member is given the opportunity to talk extensively, often repetitiously about the experience.

8-10. Severe Stress or Stress Reaction

You do not need specialized training to recognize severe stress or stress reaction that will cause problems for the service member, the unit, or the mission. Reactions that are less severe, however, are more difficult to detect. To determine whether a person needs help, you must observe him to see whether he is doing something meaningful, performing his duties, taking care of himself, behaving in an unusual fashion, or acting out of character.

8-11. Application of Psychological First Aid

The emotionally disturbed service member has built a barrier against fear. He does this for his own protection, although he is probably not aware that he is doing it. If he finds that he does not have to be afraid and that there are normal, understandable things about him, he will feel safer in dropping this barrier. Persistent efforts to make him realize that you want to understand him will be reassuring, especially if you remain calm. Nothing can cause an emotionally disturbed person to become even more fearful than feeling that others are afraid of him. Try to remain calm. Familiar things, such as a cup of coffee, the use of his name, attention to a minor wound, being given a simple job to do, or the sight of familiar people and activities, will add to his ability to overcome his fear. He may not respond well if you get excited, angry, or abrupt.

a. *Ventilation.* After the service member becomes calmer, he is likely to have dreams about the stressful event. He also may think about it when he is awake or even repeat his personal reaction to the event. One benefit of this natural pattern is that it helps him master the stress by going over it just as one masters the initial fear of parachuting from an aircraft by doing it over and over again. Eventually, it is difficult to remember how frightening the event was initially. In giving first aid to the emotionally disturbed service member, you should let him follow this natural pattern. Encourage him to talk. Be a good listener. Let him tell, in his own words, what actually happened. If home front problems or worries have contributed to the stress, it will help him to talk about them. Your patient listening will prove to him that you are interested in him, and by describing his personal problem, he can work at mastering his fear. If he becomes overwhelmed in the telling, suggest a cup of coffee or a break. Whatever you do, assure him that you will listen again as soon as he is ready. Do try to help put the service member's perception of what happened back into realistic perspective; but DO NOT argue about it.

b. *Activity.*

(1) A person who is emotionally disturbed as the result of a combat action is a casualty of anxiety and fear. He is disabled because he has become temporarily overwhelmed by his anxiety. A good way to control fear is through activity. Almost all service members, for example, experience a considerable sense of anxiety and fear while they are poised, awaiting the opening of a big offensive; but this is normally relieved, and they actually feel better once they begin to move into action. They take pride in effective performance and pleasure in knowing that they are good service members, perhaps being completely unaware that overcoming their initial fear was their first major accomplishment.

(2) Useful activity is very beneficial to the emotionally disturbed service member who is not physically incapacitated. After you help a service member get over his initial fear, help him to regain some self-confidence. Make him realize his job is continuing by finding him something useful to do. Encourage him to be active. Get him to help load trucks, clean up debris, or dig fighting positions. If possible, get him back to his usual duty. Seek out his strong points and help him apply them. Avoid having him just sit around. You may have to provide direction by telling him what to do and where to do it. The instructions should be clear and simple and should be repeated. A person who has panicked is likely to argue. Respect his feelings, but point out more immediate, obtainable, and demanding needs. Channel his excessive energy and, above all, DO NOT argue. If you cannot get him interested in doing more profitable work, it may be necessary to enlist aid in controlling his overactivity before it spreads to the group and

results in more panic. Prevent the spread of such infectious feelings by restraining and segregating if necessary.

(3) Involvement in activity helps a service member in three ways; he—

- Forgets himself.
- Has an outlet for his excessive tensions.
- Proves to himself he can do something useful.

c. *Rest.* There are times, particularly in combat, when physical exhaustion is a principal cause for emotional reactions. A unit sleep plan should be established and implemented. When possible, service members should be given a safe and relatively comfortable area in which to sleep. Examples would be an area away from heavy traffic, noise, and congestion or a place that is clean and dry and protected from environmental conditions. The more uninterrupted sleep a service member gets the better he will be able to function in the tactical environment.

d. *Hygiene.* Field hygiene is an important ingredient in a service member's morale. A service member who is dirty and unkempt will not function as well as a service member who has had the opportunity to bathe and put on clean, dry clothing. During combat, unit leaders should stress the importance of personal hygiene. Good personal hygiene not only improves morale, it also is a preventive measure against disease and nonbattle injury (DNBI).

e. *Group Activity.* You have probably already noticed that a person works, faces danger, and handles serious problems better if he is a member of a closely-knit group. Each service member in the team supports the other team members. Esprit de corps is built because the service members have the same interests, goals, and mission, and as a result they are more productive; furthermore, they are less worried because everyone is involved. It is this spirit that takes a strategic hill in battle. It is so powerful that it is one of the most effective tools you have in your *psychological first aid bag*. Getting the service member back into the team or squad activities will reestablish his sense of belonging and security and will go far toward making him a useful member of the unit.

8-12. Reactions and Limitations

Up to this point the discussion has been primarily about the feelings of the emotionally distressed service member. What about your feelings toward him? Whatever the situation, you will have emotional reactions (conscious or unconscious) toward this service member. Your reactions can either help or hinder your ability to help him. When you are tired or worried, you may very easily become impatient with him if he is unusually slow or exaggerates. You may even feel resentful toward him. At times when many physically wounded lie about you, it will be especially natural for you to resent disabilities that you cannot see. Physical wounds can be seen and easily accepted. Emotional reactions are more difficult to accept as injuries. On the other hand, will you tend to be overly sympathetic? Excessive sympathy for an incapacitated person can be as harmful as negative feelings in your relationship with him. He needs strong help, but not your sorrow. To overwhelm him with pity will make him feel even more inadequate. You must expect your buddy to recover, to be able to return to duty, and to become a useful service member again. This expectation should be displayed in your behavior and attitude as well as in what you say. If he can see your calmness, confidence, and competence, he will be reassured and will feel a sense of greater security.

8-13. Stress Reactions

See Tables 8-1, 8-2, and 8-3 for more information.

Table 8-1. Mild Stress Reaction

PHYSICAL SIGNS*	EMOTIONAL SIGNS*
1. TREMBLING, TEARFUL 2. JUMPINESS, NERVOUSNESS 3. COLD SWEAT, DRY MOUTH 4. POUNDING HEART, DIZZINESS 5. INSOMNIA, NIGHTMARES 6. NAUSEA, VOMITING, DIARRHEA 7. FATIGUE 8. THOUSAND-YARD STARE 9. DIFFICULTY THINKING, SPEAKING, AND COMMUNICATING	1. ANXIETY, INDECISIVENESS 2. IRRITABLE, COMPLAINING 3. FORGETFUL, UNABLE TO CONCENTRATE 4. EASILY STARTLED BY NOISE, MOVEMENT 5. GRIEF, TEARFUL 6. ANGER, BEGINNING TO LOSE CONFIDENCE IN SELF AND UNIT

SELF- AND BUDDY AID
1. CONTINUE MISSION PERFORMANCE, FOCUS ON IMMEDIATE MISSION. 2. EXPECT SERVICE MEMBER TO PERFORM ASSIGNED DUTIES. 3. REMAIN CALM AT ALL TIMES; BE DIRECTIVE AND IN CONTROL. 4. LET SERVICE MEMBER KNOW HIS REACTION IS NORMAL, AND THAT THERE IS NOTHING SERIOUSLY WRONG WITH HIM. 5. KEEP SERVICE MEMBER INFORMED OF THE SITUATION, OBJECTIVES, EXPECTATIONS, AND SUPPORT. CONTROL RUMORS. 6. BUILD SERVICE MEMBER'S CONFIDENCE, TALK ABOUT SUCCEEDING. 7. KEEP SERVICE MEMBER PRODUCTIVE (WHEN NOT RESTING) THROUGH RECREATIONAL ACTIVITIES, EQUIPMENT MAINTENANCE.

8. ENSURE SERVICE MEMBER MAINTAINS GOOD PERSONAL HYGIENE.
9. ENSURE SERVICE MEMBER EATS, DRINKS, AND SLEEPS AS SOON AS POSSIBLE.
10. LET SERVICE MEMBER TALK ABOUT HIS FEELINGS. DO NOT "PUT DOWN" HIS FEELINGS OF GRIEF OR WORRY. GIVE PRACTICAL ADVICE AND PUT EMOTIONS INTO PERSPECTIVE.

* MOST OR ALL OF THESE SIGNS ARE PRESENT IN MILD STRESS REACTION. THEY CAN BE PRESENT IN ANY NORMAL SERVICE MEMBER IN COMBAT YET HE CAN STILL DO HIS JOB.

Table 8-2. More Serious Stress Reaction

PHYSICAL SIGNS*	EMOTIONAL SIGNS*
1. CONSTANTLY MOVES AROUND	1. RAPID AND/OR INAPPROPRIATE TALKING
2. FLINCHING OR DUCKING AT SUDDEN SOUNDS	2. ARGUMENTATIVE, RECKLESS MOVEMENTS/ACTIONS
3. SHAKING, TREMBLING (WHOLE BODY OR ARMS)	3. INATTENTIVE TO PERSONAL HYGIENE
4. CANNOT USE PART OF BODY, NO PHYSICAL REASON (HAND, ARM, LEGS)	4. INDIFFERENT TO DANGER
5. CANNOT SEE, HEAR, OR FEEL (PARTIAL OR COMPLETE LOSS)	5. MEMORY LOSS
6. PHYSICAL EXHAUSTION, CRYING	6. SEVERE STUTTERING, MUMBLING, OR CANNOT SPEAK AT ALL
7. FREEZING UNDER FIRE, OR TOTAL IMMOBILITY	7. INSOMNIA, NIGHTMARES
8. VACANT STARES, STAGGERS, SWAYS WHEN STANDS	8. SEEING OR HEARING THINGS THAT DO NOT EXIST
9. PANIC RUNNING UNDER FIRE	9. RAPID EMOTIONAL SHIFTS
	10. SOCIAL WITHDRAWAL
	11. APATHETIC
	12. HYSTERICAL OUTBURSTS
	13. FRANTIC OR STRANGE BEHAVIOR

TREATMENT PROCEDURES**

1. IF A SERVICE MEMBER'S BEHAVIOR ENDANGERS THE MISSION, SELF, OR OTHERS, DO WHATEVER IS NECESSARY TO CONTROL HIM.
2. IF THE SERVICE MEMBER IS UPSET, CALMLY TALK HIM INTO COOPERATING.
3. IF CONCERNED ABOUT THE SERVICE MEMBER'S RELIABILITY:
   - UNLOAD HIS WEAPON.
   - TAKE WEAPON IF SERIOUSLY CONCERNED.
   - PHYSICALLY RESTRAIN HIM ONLY WHEN NECESSARY FOR SAFETY OR TRANSPORTATION.
4. REASSURE EVERYONE THAT THE SIGNS ARE PROBABLY JUST STRESS REACTION AND WILL QUICKLY IMPROVE.
5. IF STRESS REACTION SIGNS CONTINUE:
   - GET THE SERVICE MEMBER TO A SAFER PLACE.
   - DO NOT LEAVE THE SERVICE MEMBER ALONE, KEEP SOMEONE HE KNOWS WITH HIM.
   - NOTIFY SENIOR NONCOMMISSIONED OFFICER (NCO) OR OFFICER.
   - HAVE THE SERVICE MEMBER EXAMINED BY MEDICAL PERSONNEL.

Table 8-2. More Serious Stress Reaction (Continued)

TREATMENT PROCEDURES**
6. GIVE THE SERVICE MEMBER EASY TASKS TO DO WHEN NOT SLEEPING, EATING, OR RESTING. 7. ASSURE THE SERVICE MEMBER HE WILL RETURN TO FULL DUTY IN 24 HOURS; AND, RETURN HIM TO NORMAL DUTIES AS SOON AS HE IS READY.
* THESE SIGNS ARE PRESENT IN ADDITION TO THE SIGNS OF MILD STRESS REACTION. ** DO THESE PROCEDURES IN ADDITION TO THE SELF- AND BUDDY AID CARE.

Table 8-3. Preventive Measures to Combat Stress Reaction

1. WELCOME NEW MEMBERS INTO YOUR TEAM, GET TO KNOW THEM QUICKLY. IF YOU ARE NEW, BE ACTIVE IN MAKING FRIENDS.
2. BE PHYSICALLY FIT (STRENGTH, ENDURANCE, AND AGILITY).
3. KNOW AND PRACTICE LIFESAVING SELF- AND BUDDY AID.
4. PRACTICE RAPID RELAXATION TECHNIQUES (FM 22-51).
5. HELP EACH OTHER OUT WHEN THINGS ARE TOUGH AT HOME OR IN THE UNIT.
6. KEEP INFORMED; ASK YOUR LEADER QUESTIONS, IGNORE RUMORS.
7. WORK TOGETHER TO GIVE EVERYONE FOOD, WATER, SHELTER, HYGIENE, AND SANITATION.
8. SLEEP WHEN MISSION AND SAFETY PERMIT; LET EVERYONE GET TIME TO SLEEP.
   - SLEEP ONLY IN SAFE PLACES AND BY STANDING OPERATING PROCEDURE (SOP).
   - IF POSSIBLE, SLEEP 6 TO 9 HOURS PER DAY.
   - TRY TO GET AT LEAST 4 HOURS SLEEP PER DAY.
   - GET GOOD SLEEP BEFORE GOING ON SUSTAINED OPERATIONS.
   - CATNAP WHEN YOU CAN, BUT ALLOW TIME TO WAKE UP FULLY.
   - CATCH UP ON SLEEP AFTER GOING WITHOUT.

APPENDIX A

FIRST AID CASE AND KITS, DRESSINGS, AND BANDAGES

A-1. First Aid Case with Field Dressings and Bandages

Every service member is issued a first aid case (Figure A-1A) with a field first aid dressing encased in a plastic wrapper (Figure A-1B). He carries it at all times for his use. The field first aid dressing is a standard sterile (germ-free) compress or pad with bandages attached (Figure A-1C). This dressing is used to cover the wound, to protect against further contamination, and to stop bleeding (pressure dressing). When a service member administers first aid to another person, he must remember to use the wounded person's dressing; he may need his own later. The service member must check his first aid case regularly and replace any used or missing dressing. The field first aid dressing may normally be obtained from his unit supply.

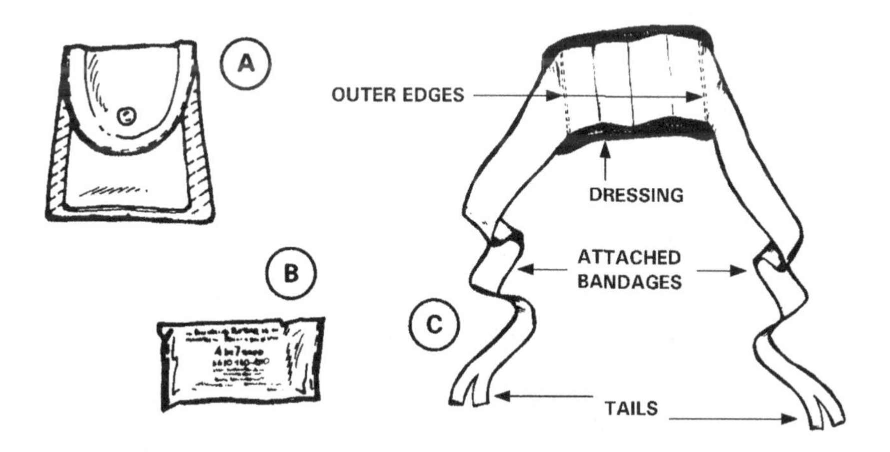

Figure A-1. Field first aid case and dressing (Illustrated A—C).

A-2. General Purpose First Aid Kits

General-purpose first aid kits are listed in the common table of allowances (CTA) 8-100. The operators, crew, and passengers carry these kits on Army vehicles, aircraft, and boats for use. Individuals designated by unit TSOP to be responsible for these kits are required to check them regularly and replace all items used. The general-purpose kit and its contents can be obtained through the unit supply system.

NOTE

Periodically check the dressings (for holes or tears in the packaging) and the medicines (for expiration date) that are in the first aid kits. If necessary, replace defective or outdated items.

A-3. Dressings

Dressings are sterile pads or compresses used to cover wounds. They usually are made of gauze or cotton wrapped in gauze (Figure A-1C). In addition to the standard field first aid dressing, other dressings such as sterile gauze compresses and small sterile compresses on adhesive strips may be available under CTA 8-100.

A-4. Standard Bandages

a. Standard bandages are made of gauze or muslin and are used over a sterile dressing to secure the dressing in place, to close off its edge from dirt and germs, and to create pressure on the wound and control bleeding. A bandage can also support an injured part or secure a splint.

b. Tailed bandages may be attached to the dressing as indicated on the field first aid dressing (Figure A-1C).

A-5. Triangular and Cravat (Swathe) Bandages

a. Triangular and cravat (or swathe) bandages (Figure A-2) are fashioned from a triangular piece of muslin (37 by 37 by 52 inches) provided in the general-purpose first aid kit. If it is folded into a strip, it is called a cravat. Two safety pins are packaged with each bandage. These bandages are valuable in an emergency since they are easily applied.

b. To improvise a triangular bandage, cut a square of available material, slightly larger than 3 feet by 3 feet, and *fold it diagonally*. If two bandages are needed, cut the material along the diagonal fold.

c. A cravat can be improvised from such common items as T-shirts, other shirts, bed linens, trouser legs, scarfs, or any other item made of pliable and durable material that can be folded, torn, or cut to the desired size.

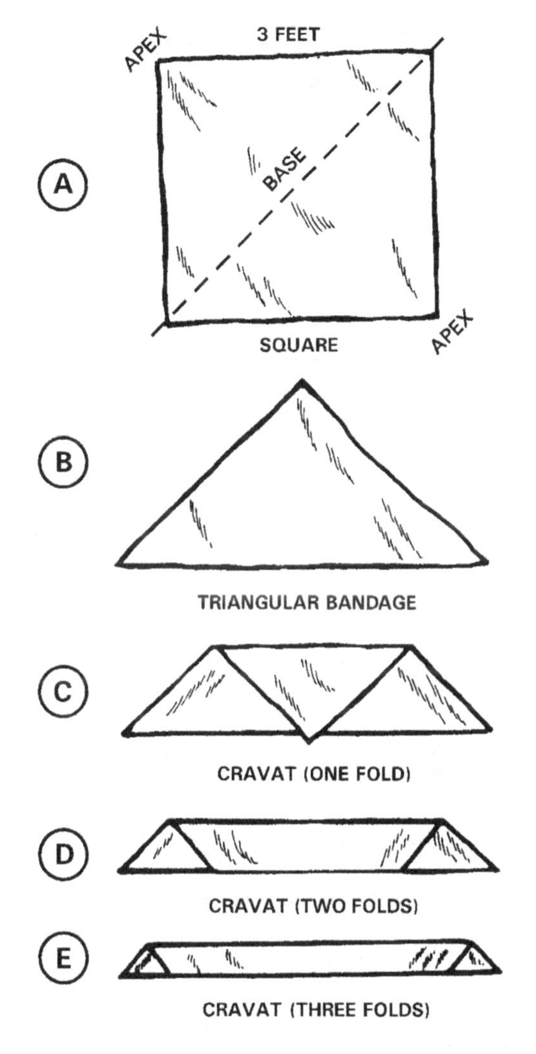

Figure A-2. Triangular and cravat bandages (Illustrated A—E).

APPENDIX B

RESCUE AND TRANSPORTATION PROCEDURES

B-1. General

A basic principle of first aid is to evaluate the casualty's injuries and administer first aid before moving him. However, adverse situations or conditions may jeopardize the lives of both the rescuer and the casualty if this is done. It may be necessary *first to rescue* the casualty before first aid can be effectively or safely given. The life and/or the well-being of the casualty will depend as much upon the manner in which he is *rescued and transported,* as it will upon the first aid and medical treatment he receives. Rescue actions must be done quickly and safely. Careless or rough handling of the casualty during rescue operations can aggravate his injuries.

B-2. Principles of Rescue Operations

a. When faced with the necessity of rescuing a casualty who is threatened by hostile action, fire, water, or any other immediate hazard, DO NOT take action without first determining the extent of the hazard and your ability to handle the situation. DO NOT become a casualty.

b. The rescuer must evaluate the situation and analyze the factors involved. This evaluation involves three major steps:

- Identify the task.
- Evaluate circumstances of the rescue.
- Plan the action.

B-3. Considerations

a. First determine if a rescue attempt is actually needed. It is a waste of time, equipment, and personnel to rescue someone not in need of rescuing. It is also a waste to look for someone who is not lost or needlessly risk the lives of the rescuer(s). In planning a rescue, attempt to obtain the following information:

- Who, what, where, when, why, and how the situation happened?

- How many casualties are involved and the nature of their injuries?

- What is the tactical situation?

- What are the terrain features and the location of the casualties?

- Will there be adequate assistance available to aid in the rescue/evacuation?

- Can first aid and/or medical treatment be provided at the scene; will the casualties require movement to a safer location?

- What specialized equipment will be required for the rescue operation?

- Is the rescue area contaminated? Will decontamination equipment and materiel be required for casualties, rescue personnel, and rescue equipment?

- How much time is available?

b. The time element can play a significant role in how the rescue is attempted. If the casualties are in imminent danger of losing their lives (such as near a burning vehicle or in a burning building) the time available will be relatively short and will sometimes cause a rescuer to compromise planning stages and/or the first aid which can be given. However, if the casualty is in a relatively secure area and his physical condition is strong, more deliberate planning can take place. A realistic estimate of time available must be made as quickly as possible to determine action time remaining. The key elements are the casualty's physical and mental condition, the tactical situation, and the environment.

B-4. Plan of Action

a. The casualty's ability to endure is of primary importance in estimating the time available. Age, physical condition, and extent of wounds and/or injuries will differ from casualty to casualty. Therefore, to determine the time available, you will have to consider—

- Endurance time of the casualty.

- Extent of injuries.

- Type of situation.
- Personnel and/or equipment availability.
- Weather.
- Terrain (natural and man-made).
- Environment (contaminated or uncontaminated).

b. In respect to terrain, you must consider altitude and visibility. In some cases, the casualty may be of assistance because he knows more about the particular terrain or situation than you do. Maximum use of secure/reliable trails or roads is essential.

c. When taking weather into account, ensure that blankets and/or rain gear are available. Even a mild rain can complicate a normally simple rescue. In high altitudes and/or extreme cold and gusting winds, the time available is critically shortened. Be prepared to provide shelter and warmth for the casualty as well as the rescuers.

B-5. Proper Handling of Casualties

a. You may have saved the casualty's life through the application of appropriate first aid measures. However, his life can be lost through rough handling or careless transportation procedures. Before you attempt to move the casualty—

- Evaluate the type and extent of his injuries.
- Ensure that dressings over wounds are adequately reinforced.
- Ensure that fractured bones are properly immobilized and supported to prevent them from cutting through muscle, blood vessels, and skin.

b. Based upon your evaluation of the type and extent of the casualty's injury and your knowledge of the various manual carries, you must select the best possible method of manual transportation. If the casualty is conscious, tell him how he is to be transported. This will help allay his fear of movement and gain his cooperation and confidence.

c. Buddy aid for chemical agent casualties includes those actions required to prevent an incapacitated casualty from receiving additional injury from the effects of chemical hazards. If a casualty is physically unable to decontaminate himself or administer the proper chemical agent antidote, the casualty's buddy assists him and assumes responsibility for his care. Buddy-aid includes—

- Administering the proper chemical agent antidote.
- Decontaminating the incapacitated casualty's exposed skin.
- Ensuring that his protective ensemble remains correctly emplaced.
- Maintaining respiration.
- Controlling bleeding.
- Providing other standard first aid measures
- Transporting the casualty out of the contaminated area.

B-6. Positioning the Casualty

The first step in any manual carry is to position the casualty to be lifted. If he is conscious, he should be told how he is to be positioned and transported. This helps lessen his fear of movement and to gain his cooperation. It may be necessary to roll the casualty onto his abdomen, or his back, depending upon the position in which he is lying and the particular carry to be used.

a. To roll a casualty onto his abdomen, kneel at the casualty's uninjured side.

(1) Place his arms above his head; cross his ankle which is farther from you over the one that is closer to you.

(2) Place your hands on the shoulder which is farther from you; place your other hand in the area of his hip or thigh (Figure B-1).

(3) Roll him gently toward you onto his abdomen (Figure B-2).

b. To roll a casualty onto his back, follow the same procedure described in *a* above, except gently roll the casualty onto his back, rather than onto his abdomen.

Figure B-1. Positioning the casualty.

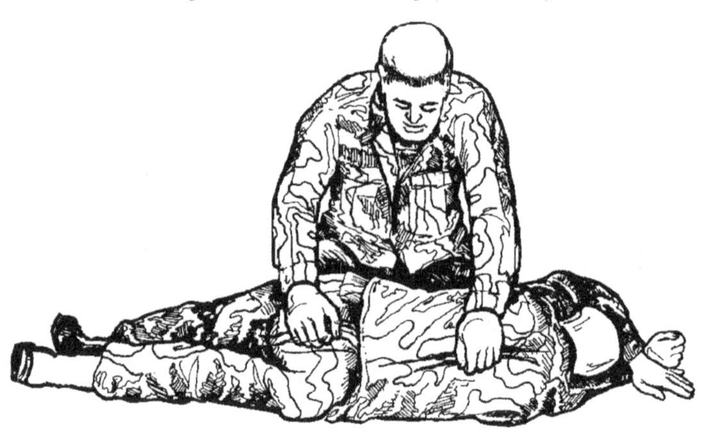

Figure B-2. Rolling casualty onto his abdomen.

B-7. Medical Evacuation and Transportation of Casualties

a. Medical evacuation of the sick and wounded (with en route medical care) is the responsibility of medical personnel who have been provided special training and equipment. Therefore, unless a good reason for you to transport a casualty arises, wait for some means of medical evacuation to be provided. When the situation is urgent and you are unable to obtain medical assistance or know that no medical evacuation assets are available, you will have to transport the casualty. For this reason, you must know how to transport him without increasing the seriousness of his condition.

b. Transporting a casualty by litter (FM 8-10-6) is safer and more comfortable for him than by manual means; it is also easier for you.

Manual transportation, however, may be the only feasible method because of the terrain or the combat situation; or it may be necessary to save a life. In these situations, the casualty should be transferred to a litter as soon as one can be made available or improvised.

B-8. Manual Carries

Casualties carried by manual means must be carefully and correctly handled, otherwise their injuries may become more serious or possibly fatal. Situation permitting, transport of a casualty should be organized and unhurried. Each movement should be performed as deliberately and gently as possible. Casualties should not be moved before the type and extent of injuries are evaluated and the required first aid is administered. The exception to this occurs when the situation dictates immediate movement for safety purposes (for example, it may be necessary to remove a casualty from a burning vehicle); that is, the situation dictates that the urgency of casualty movement outweighs the need to administer first aid. Manual carries are tiring for the bearers and involve the risk of increasing the severity of the casualty's injury. In some instances, however, they are essential to save the casualty's life. Although manual carries are accomplished by one or two bearers, the two-man carries are used whenever possible. They provide more comfort to the casualty, are less likely to aggravate his injuries, and are also less tiring for the bearers. The distance a casualty can be carried depends on many factors, such as—

- Nature of the casualty's injuries.
- Strength and endurance of the bearer(s).
- Weight of the casualty.
- Obstacles encountered during transport (natural or manmade).
- Type of terrain.

a. One-man Carries. These carries should be used when only one bearer is available to transport the casualty.

(1) The *fireman's carry* (Figure B-3) is one of the easiest ways for one individual to carry another. After an unconscious or disabled casualty has been properly positioned, he is raised from the ground, then supported and placed in the carrying position.

(*a*) After rolling the casualty onto his abdomen, straddle him. Extend your hands under his chest and lock them together.

(*b*) Lift the casualty to his knees as you move backward.

(*c*) Continue to move backward, thus straightening the casualty's legs and locking his knees.

(*d*) Walk forward, bringing the casualty to a standing position; tilt him slightly backward to prevent his knees from buckling.

(*e*) As you maintain constant support of the casualty with one arm, free your other arm, quickly grasp his wrist, and raise his arm high. Instantly pass your head under his raised arm, releasing it as you pass under it.

(*f*) Move swiftly to face the casualty and secure your arms around his waist. Immediately place your foot between his feet and spread them apart (approximately 6 to 8 inches).

(*g*) Grasp the casualty's wrist and raise his arm high over your head.

(*h*) Bend down and pull the casualty's arm over and down on your shoulder, bringing his body across your shoulders. At the same time, pass your arm between his legs.

(*i*) Grasp the casualty's wrist with one hand, and place your other hand on your knee for support.

(*j*) Rise with the casualty positioned correctly. Your other hand is free for use.

Figure B-3. Fireman's carry (Illustrated A—J).

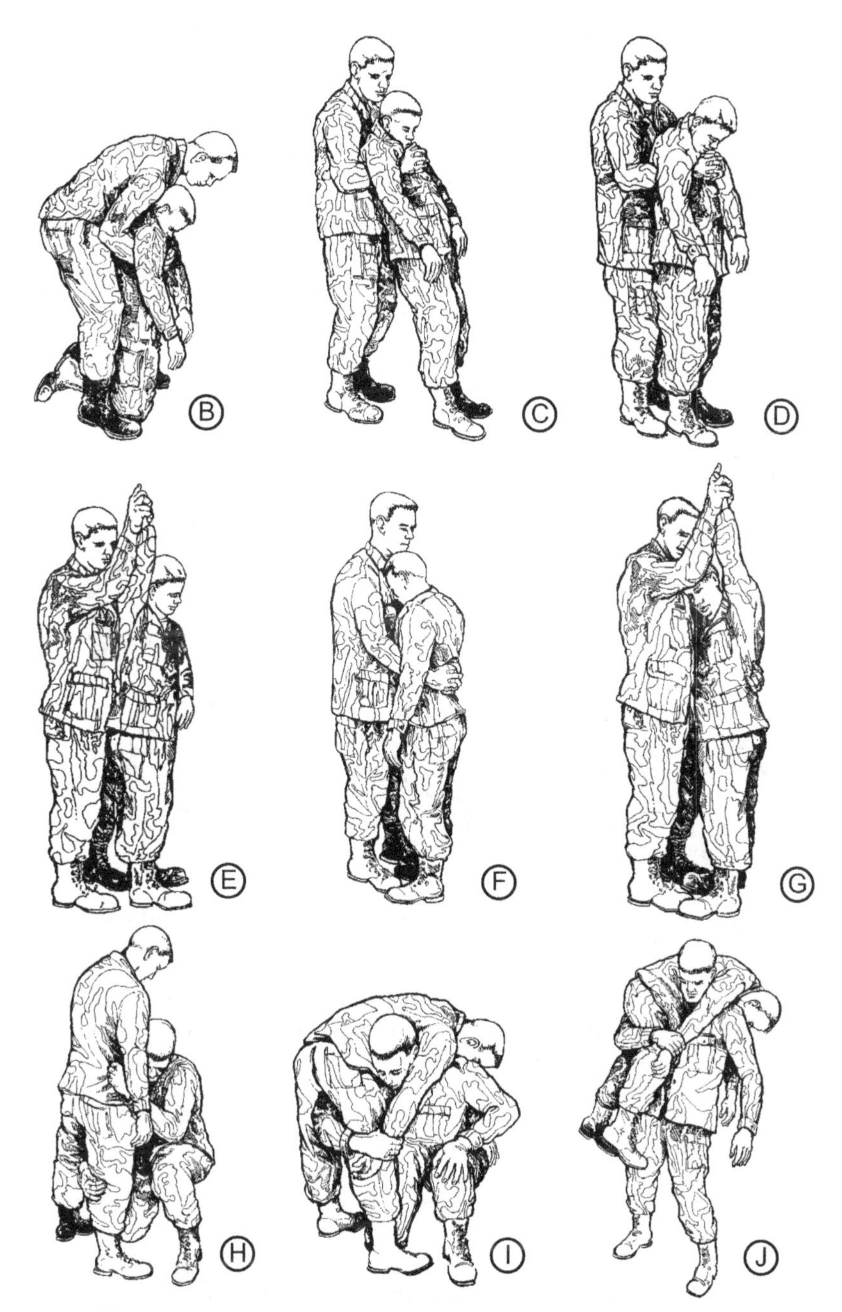

Figure B-3. Fireman's carry (Illustrated A—J) (Continued).

(2) The alternate method of the *fireman's carry* for raising a casualty from the ground is illustrated in Figure B-4; however, it should be used only when the bearer believes it to be safer for the casualty because of the location of his wounds. When the alternate method is used, care must be taken to prevent the casualty's head from snapping back and causing a neck injury. The steps for raising a casualty from the ground for the fireman's carry are also used in other one-man carries.

(*a*) Kneel on one knee at the casualty's head and face his feet. Extend your hands under his armpits, down his sides, and across his back.

(*b*) As you rise, lift the casualty to his knees. Then secure a lower hold and raise him to a standing position with his knees locked.

Figure B-4. Fireman's carry (alternate method) for lifting a casualty to a standing position (Illustrated A—B).

(3) In the *supporting carry* (Figure B-5), the casualty must be able to walk or at least hop on one leg, using the bearer as a crutch. This carry can be used to assist him as far as he is able to walk or hop.

(*a*) Raise the casualty from the ground to a standing position by using the fireman's carry.

(*b*) Grasp the casualty's wrist and draw his arm around your neck.

(*c*) Place your arm around his waist. The casualty is now able to walk or hop using you as a support.

Figure B-5. Supporting carry.

(4) The *arms carry* (Figure B-6) is useful in carrying a casualty for a short distance (up to 50 meters) and for placing him on a litter.

(*a*) Raise or lift the casualty from the ground to a standing position, as in the fireman's carry.

(*b*) Place one arm under the casualty's knees and your other arm around his back.

(*c*) Lift the casualty.

(*d*) Carry the casualty high to lessen fatigue.

Figure B-6. Arms carry.

(5) Only a conscious casualty can be transported by the *saddleback carry* (Figure B-7), because he must be able to hold onto the bearer's neck. To use this technique—

(*a*) Raise the casualty to an upright position, as in the fireman's carry.

(*b*) Support the casualty by placing an arm around his waist. Move to the casualty's side. Have the casualty put his arm around your neck and move in front of him with your back to support him.

(*c*) Have the casualty encircle his arms around your neck

(*d*) Stoop, raise him on your back and clasp your hands together beneath his thighs, if possible.

Figure B-7. Saddleback carry.

(6) In the *pack-strap carry* (Figure B-8), the casualty's weight rests high on the your back. This makes it easier for you to carry the casualty a moderate distance (50 to 300 meters). To eliminate the possibility of injury to the casualty's arms, you must hold his arms in a palms-down position.

(*a*) Lift the casualty from the ground to a standing position, as in the fireman's carry.

(*b*) Support the casualty with your arms around him and grasp his wrist closer to you.

(*c*) Place his arm over your head and across your shoulders.

(*d*) Move in front of him while still supporting his weight against your back.

(*e*) Grasp his other wrist and place this arm over your shoulder.

(*f*) Bend forward and raise or hoist the casualty as high on your back as possible so that his weight is resting on your back.

NOTE

Once the casualty is positioned on the bearer's back, the bearer remains as erect as possible to prevent straining or injuring his back.

Figure B-8. Pack-strap carry.

(7) The *pistol-belt carry* (Figure B-9) is the best one-man carry for a long distance (over 300 meters). The casualty is securely supported upon your shoulders by a belt. Both your hands and the casualty's (if conscious) are free for carrying a weapon or equipment, or climbing obstacles. With your hands free and the casualty secured in place, you are also able to creep through shrubs and under low-hanging branches.

(*a*) Link two pistol belts (or three, if necessary) together to form a sling. Place the sling under the casualty's thighs and lower back so that a loop extends from each side.

NOTE

If pistol belts are not available for use, other items such as a rifle sling, two cravat bandages, two litter straps, or any other suitable material, which will not cut or bind the casualty may be used.

(*b*) Lie face up between the casualty's outstretched legs. Thrust your arms through the loops and grasp his hands and trouser leg on his injured side.

(*c*) Roll toward the casualty's uninjured side onto your abdomen, bringing him onto your back. Adjust the sling, if necessary.

(*d*) Rise to a kneeling position. The belt will hold the casualty in place.

(*e*) Place one hand on your knee for support and rise to an upright position. (The casualty is supported on your shoulders.)

(*f*) Carry the casualty with your hands free for use in rifle firing, climbing, or surmounting obstacles.

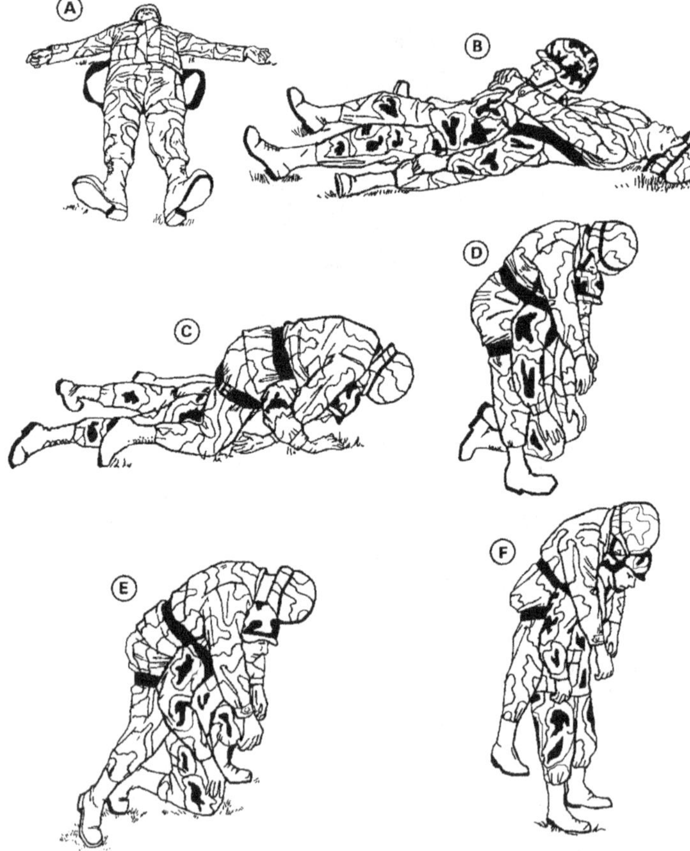

Figure B-9. Pistol-belt carry (Illustrated A—F).

(8) The *pistol-belt drag* (Figure B-10), as well as other drags, is generally used for short distances (up to 50 meters). This drag is useful in combat, since both the bearer and the casualty can remain closer to the ground than in any other drags.

(*a*) Extend two pistol belts or similar objects to their full length and join them together to make a continuous loop.

(*b*) Roll the casualty onto his back, as in the fireman's carry.

(*c*) Pass the loop over the casualty's head, and position it across his chest and under his armpits. Then cross the remaining portion of the loop, thus forming a figure eight. Keep tension on the belts so they do not come unhooked.

(*d*) Lie on your side facing the casualty.

(*e*) Slip the loop over your head and turn onto your abdomen. This enables you to drag the casualty as you crawl.

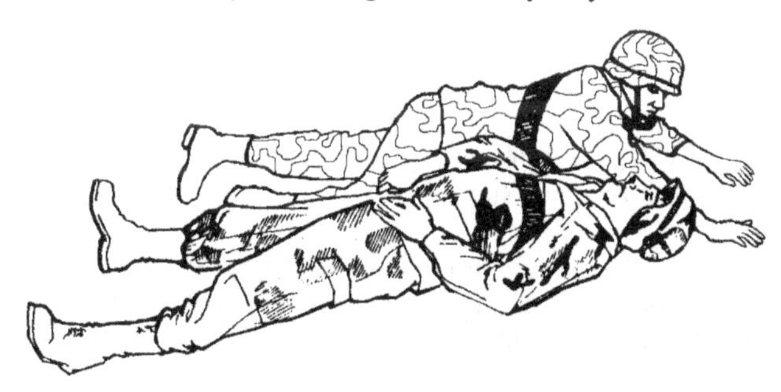

Figure B-10. Pistol-belt drag.

(9) The *neck drag* (Figure B-11) is useful in combat because the bearer can transport the casualty as he creeps behind a low wall or shrubbery, under a vehicle, or through a culvert. If the casualty is unconscious, his head must be protected from the ground. The neck drag cannot be used if the casualty has a broken arm.

NOTE

If the casualty is conscious, he may clasp his hands together around your neck.

(*a*) Tie the casualty's hands together at the wrists.

(*b*) Straddle the casualty in a kneeling face-to-face position.

(*c*) Loop the casualty's tied hands over and around your neck.

(*d*) Crawl forward dragging the casualty with you.

NOTE

If the casualty is unconscious, protect his head from the ground.

Figure B-11. Neck drag.

(10) The *cradle drop drag* (Figure B-12) is effective in moving a casualty up or down steps.

(*a*) Kneel at the casualty's head (with him lying on his back). Slide your hands, with palms up, under the casualty's shoulders and get a firm hold under his armpits.

(*b*) Rise (partially), supporting the casualty's head on one of your forearms. (You may bring your elbows together and let the casualty's head rest on both of your forearms.)

(*c*) Rise and drag the casualty backward. (The casualty is in a semisitting position.)

(*d*) Back down the steps, supporting the casualty's head and body and letting his hips and legs drop from step to step.

NOTE

If the casualty needs to be moved up the steps, you should back up the steps, using the same procedure.

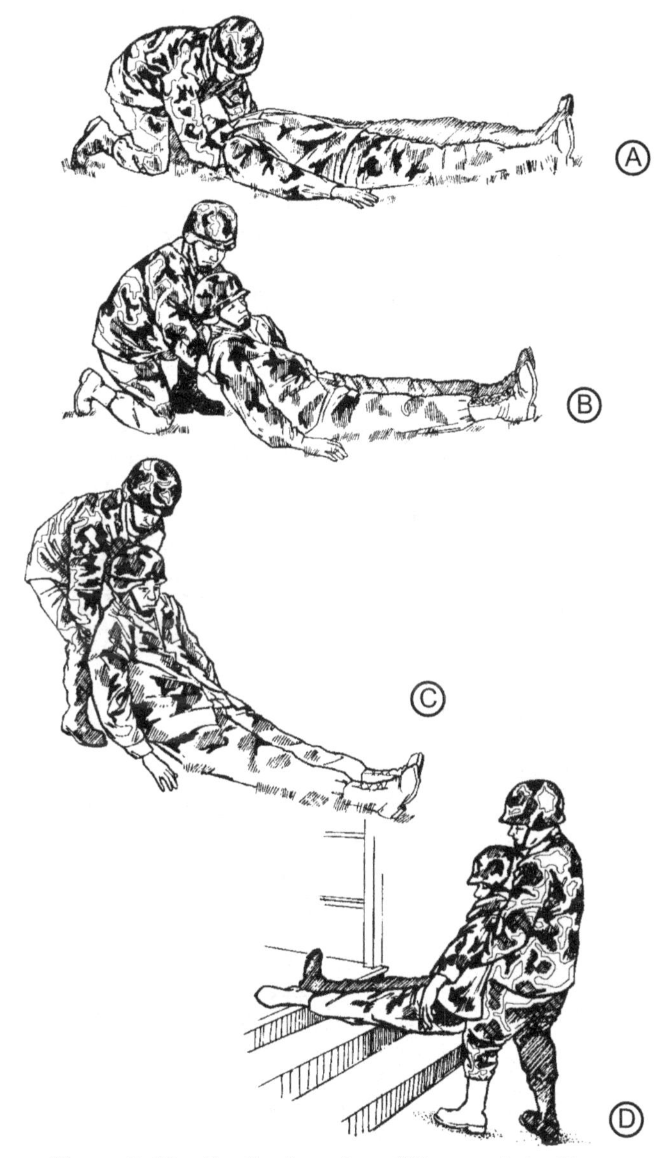

Figure B-12. Cradle-drop drag (Illustrated A—D).

(11) The *LBE carry using the bearer's LBE* can be used with a conscious casualty (Figure B-13).

(*a*) Loosen all suspenders on your LBE.

(*b*) Have the casualty place one leg into the loop formed by your suspenders and pistol belt.

(*c*) Squat in front of the standing casualty. Have him place his other leg into the loop, also.

(*d*) Have the casualty place his arms over your shoulders, lean forward onto your back, and lock his hands together.

(*e*) Stand up and lean forward into a comfortable position.

(*f*) Continue the mission.

Figure B-13. Load bearing equipment carry using bearer's LBE (conscious casualty) (Illustrated A—F).

Figure B-13. Load bearing equipment carry using bearer's LBE (conscious casualty) (Illustrated A—F) (Continued).

(12) The *LBE carry using the bearer's LBE* can be used with an unconscious casualty or one who cannot stand (Figure B-14).

(*a*) Position the casualty on the flat of his back.

(*b*) Remove your LBE and loosen all suspender straps.

(*c*) Lift the casualty's leg and place it through the loop formed by your suspenders and pistol belt. Then place the other leg through the same loop. The LBE is moved up until the pistol belt is behind the casualty's thighs.

(*d*) Lay between the casualty's legs; work your arms through the LBE suspenders.

(*e*) Grasp the casualty's hand (on the injured side), and roll the casualty (on his uninjured side) onto your back.

(*f*) Rise to one knee and then push into a standing position.

(*g*) Bring the casualty's arms over your shoulders. Grasp his hands and secure them if the casualty is unconscious. If the casualty is conscious, have him lock his hands in front if he is able to do so.

(*h*) Lean forward into a comfortable position and continue the mission.

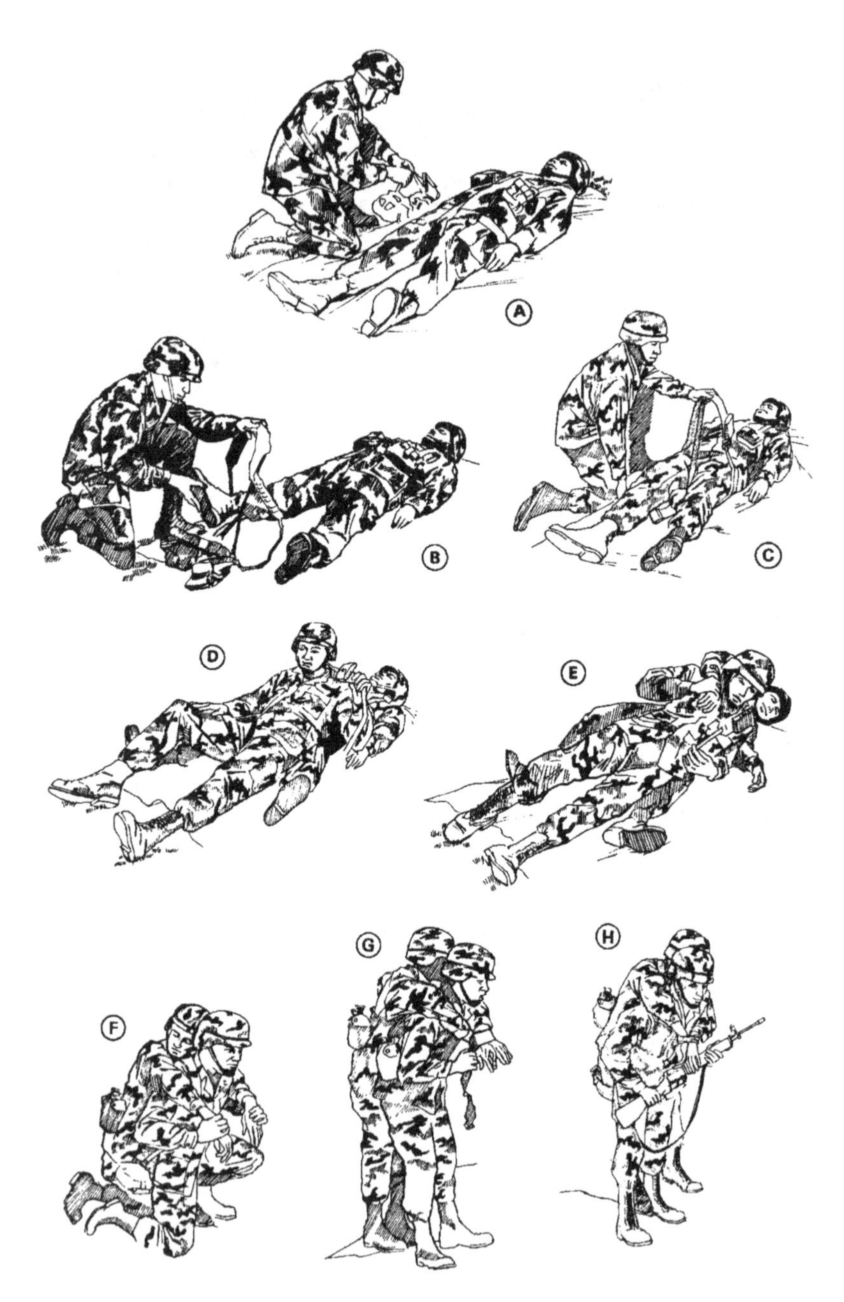

Figure B-14. Load bearing equipment carry using bearer's LBE (unconscious casualty or one that cannot stand) (Illustrated A—H).

(13) The *LBE carry using the casualty's LBE* (Figure B-15) can be used with a conscious or unconscious casualty.

(*a*) Position the casualty on his back with his LBE on.

(*b*) Loosen the casualty's two front suspenders.

(*c*) Position yourself between the casualty's legs, and slip your arms into the casualty's two front suspenders (up to your shoulders).

(*d*) Work his arms out of his LBE suspenders.

(*e*) Grasp the casualty's hand (on the injured side), and roll him (on his uninjured side) onto your back.

(*f*) Rise to one knee, then into a standing position.

(*g*) Grasp the casualty's hands and secure them, if the casualty is unconscious. Have the casualty lock his hands in front of you, if he is conscious.

(*h*) Lean forward into a comfortable position and continue the mission.

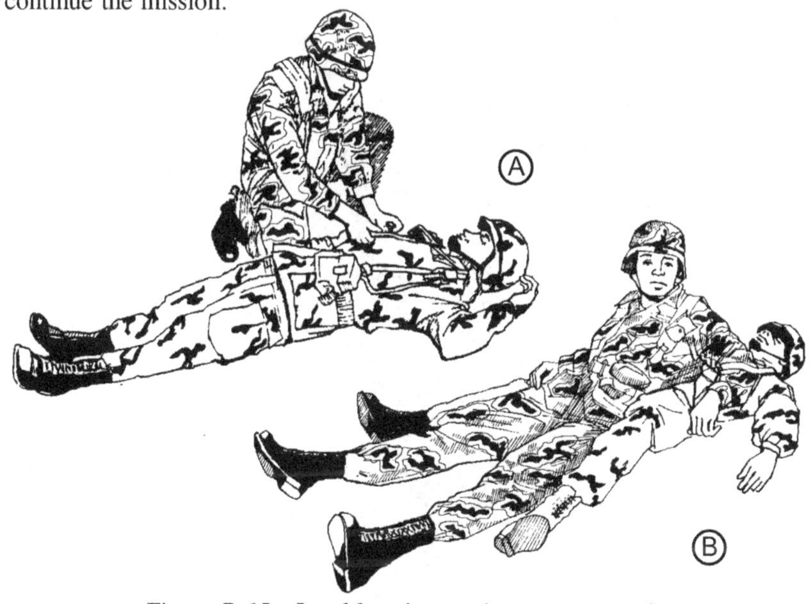

Figure B-15. Load bearing equipment carry using casualty's LBE (Illustrated A—G).

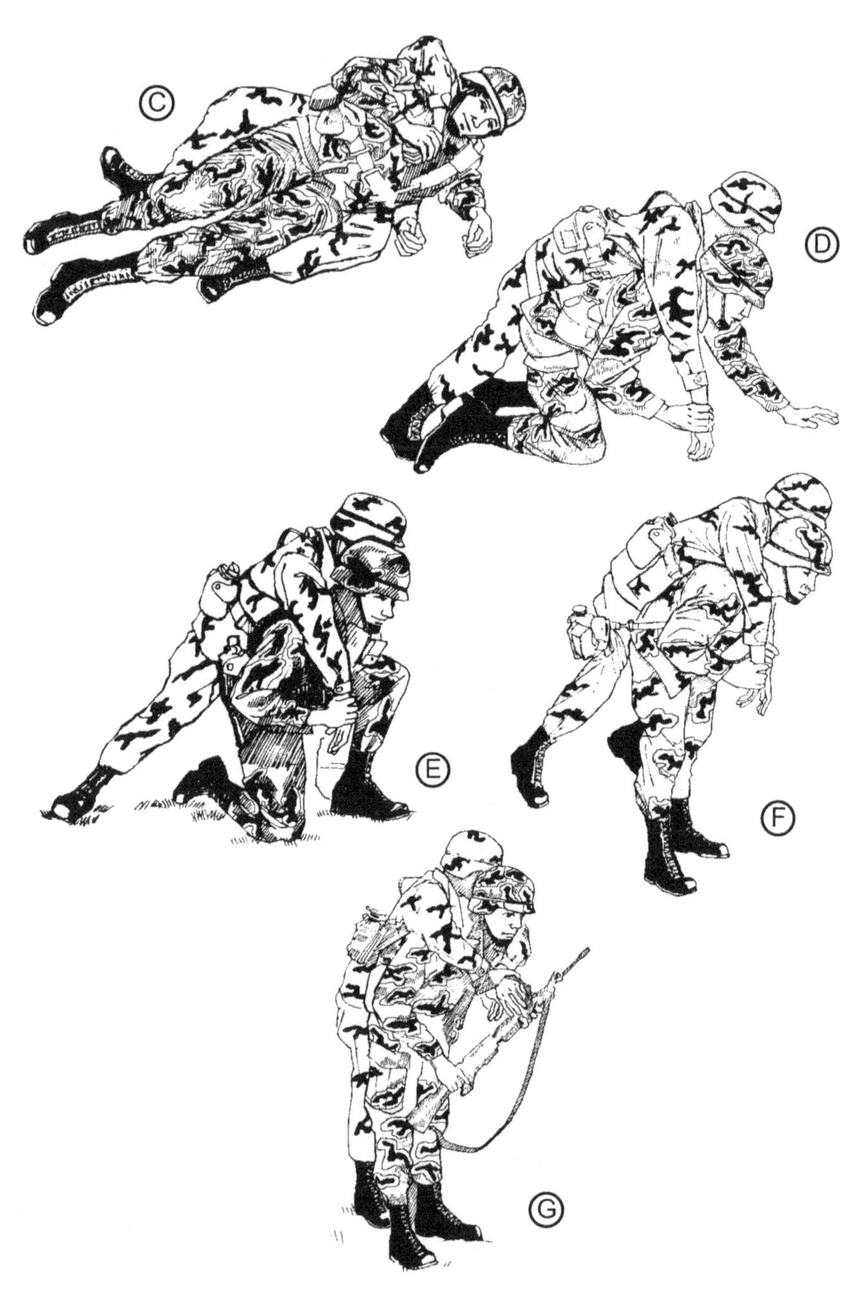

Figure B-15. Load bearing equipment carry using casualty's LBE (Illustrated A—G) (Continued).

b. *Two-man Carries*. These carries should be used whenever possible. They provide more casualty comfort, are less likely to aggravate injuries, and are less tiring for the bearers. Five different two-man carries can be used.

(1) The *two-man support carry* (Figure B-16) can be used in transporting either conscious or unconscious casualties. If the casualty is taller than the bearers, it may be necessary for the bearers to lift the casualty's legs and let them rest on their forearms. The bearers—

(*a*) Help the casualty to his feet and support him with their arms around his waist.

(*b*) Grasp the casualty's wrists and draw his arms around their necks.

Figure B-16. Two-man supporting carry.

(2) The *two-man arms carry* (Figure B-17) is useful in carrying a casualty for a moderate distance (50 to 300 meters) and placing him on a litter. To lessen fatigue, the bearers should carry the casualty high and as close to their chests as possible. In extreme emergencies when there is no time to obtain a spine board, this carry is the safest one for transporting a casualty with a back injury. If possible, two additional bearers should be used to keep the casualty's head and legs in alignment with his body. The bearers—

(*a*) Kneel at one side of the casualty; then they place their arms beneath the casualty's back, waist, hips, and knees.

(*b*) Lift the casualty while rising to their knees.

(*c*) Turn the casualty toward their chests, while rising to a standing position. Carry the casualty high to lessen fatigue.

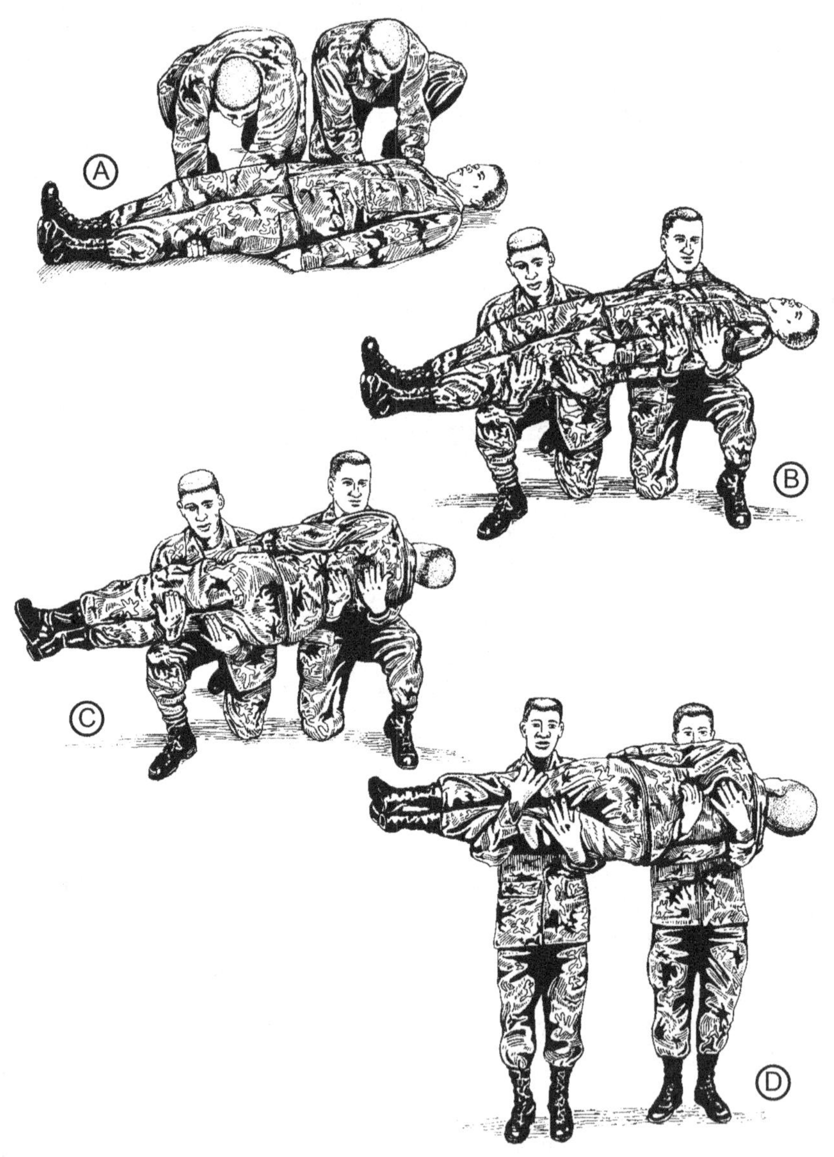

Figure B-17. Two-man arms carry (Illustrated A—D).

(3) The *two-man fore-and aft-carry* (Figure B-18) is a useful *two-man carry* for transporting a casualty for a long distance (over 300 meters). The taller of the two bearers should position himself at the casualty's head. By altering this carry so that both bearers face the casualty, it is useful for placing a casualty on a litter.

(*a*) The shorter bearer spreads the casualty's legs and kneels between them with his back to the casualty. He positions his hands behind the casualty's knees. The other bearer kneels at the casualty's head, slides his hands under the arms, across the chest, and locks his hands together.

(*b*) The two bearers rise together, lifting the casualty.

Figure B-18. Two-man fore-and-aft carry (Illustrated A—B).

(4) Only a conscious casualty can be transported with the *four-hand seat carry* (Figure B-19) because he must help support himself by placing his arms around the bearers' shoulders. This carry is especially useful in transporting a casualty with a head or foot injury for a moderate distance (50 to 300 meters). It is also useful for placing a casualty on a litter.

(*a*) Each bearer grasps one of his wrists and one of the other bearer's wrists, thus forming a packsaddle.

(*b*) The two bearers lower themselves sufficiently for the casualty to sit on the packsaddle; then, they have the casualty place his arms around their shoulders for support. The bearers then rise to an upright position.

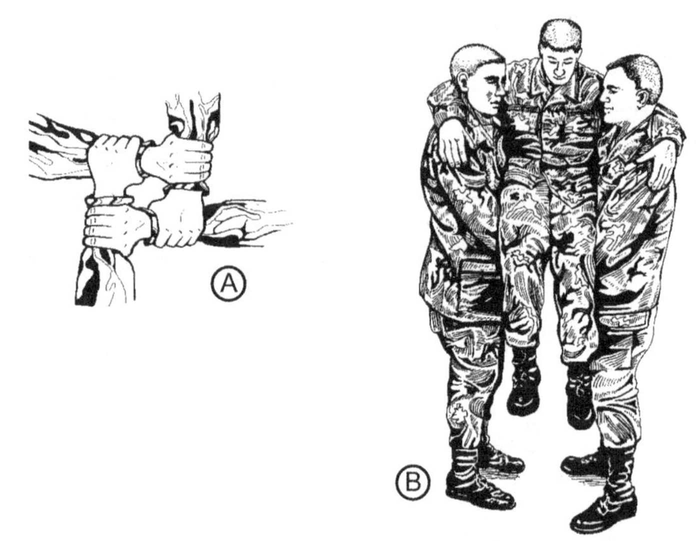

Figure B-19. Four-hand seat carry (Illustrated A—B).

(5) The *two-hand seat carry* (Figure B-20) is used when carrying a casualty for a short distance or for placing him on a litter. With the casualty lying on his back, a bearer kneels on each side of the casualty at his hips. Each bearer passes his arms under the casualty's thighs and back, and grasps the other bearer's wrists. The bearers rise lifting the casualty.

Figure B-20. Two-hand seat carry (Illustrated A—B).

B-9. Improvised Litters

Two men can support or carry a casualty without equipment for only short distances. By using available materials to improvise equipment, the casualty can be transported greater distances by two or more rescuers.

a. There are times when a casualty may have to be moved and a standard litter is not available. The distance may be too great for manual carries or the casualty may have an injury (such as a fractured neck, back, hip, or thigh) that would be aggravated by manual transportation. In these situations, litters can be improvised from materials at hand. Improvised litters must be as well constructed as possible to avoid risk of dropping or further injuring the casualty. Improvised litters are emergency measures and must be replaced by standard litters at the first opportunity.

b. Many different types of litters can be improvised, depending upon the materials available. A satisfactory litter can be made by securing poles inside such items as a blanket, poncho, shelter half, tarpaulin, mattress cover, jacket, shirt, or bed ticks, bags, and sacks (Figure B-18). Poles can be improvised from strong branches, tent supports, skis, lengths of pipe or other objects. If objects for improvising poles are not available, a blanket, poncho, or similar item can be rolled from both sides toward the center so the rolls can be gripped for carrying a patient. Most flat-surface objects of suitable size can be used as litters. Such objects include doors, boards, window shutters, benches, ladders, cots, and chairs. If possible, these objects should be padded for the casualty's comfort.

(1) To improvise a litter using a blanket and poles (Figure B-21), the following steps should be used.

Figure B-21. Litter made with blanket and poles.

(*a*) Open the blanket and lay one pole lengthwise across the center; then fold the blanket over the pole.

(*b*) Place the second pole across the center of the folded blanket.

(*c*) Fold the free edges of the blanket over the second pole and across the first pole.

(2) To improvise a litter using shirts or jackets (Figure B-22), button the shirt or jacket and turn it inside out, leaving the sleeves inside, (more than one shirt or jacket may be required), then pass the pole through the sleeves.

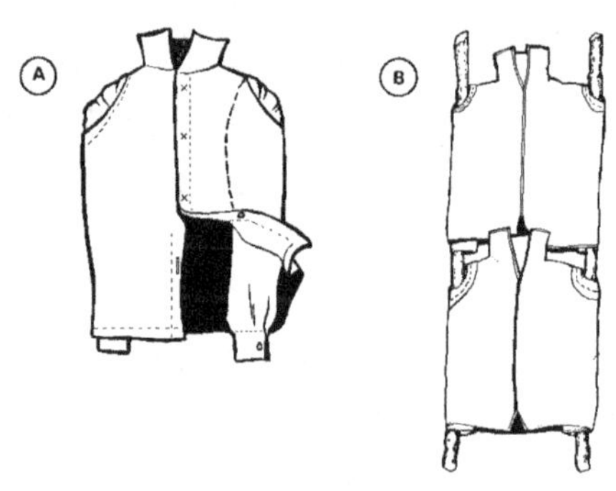

Figure B-22. Litter improvised from jackets and poles (Illustrated A—B).

(3) To improvise a litter from bed sacks and poles (Figure B-23), rip open the corners of bed ticks, bags, or sacks; then pass the poles through them.

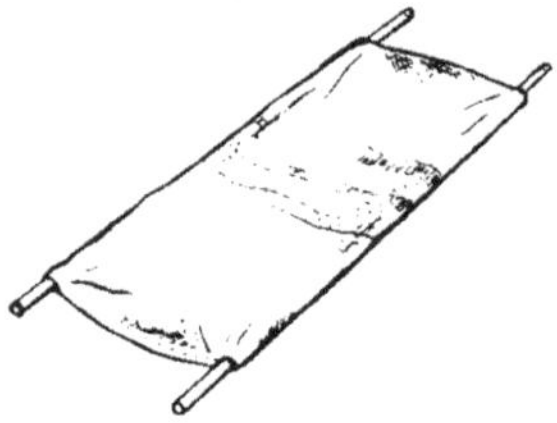

Figure B-23. Litter improvised from bed sacks and poles.

(4) If no poles are available, roll a blanket, shelter half, tarpaulin, or similar item from both sides toward the center (Figure B-24). Grip the rolls to carry the casualty.

Figure B-24. Rolled blanket used as a litter.

c. Any of the appropriate carries may be used to place a casualty on a litter. These carries are:

- The one-man arms carry (Figure B-6).
- The two-man arms carry (Figure B-17).
- The two-man fore-and-aft carry (Figure B-18).
- The two-hand seat carry (Figure B-20).
- The four-hand seat carry (Figure B-19).

WARNING

Unless there is an immediate life-threatening situation (such as fire, explosion), DO NOT move a casualty with a suspected back or neck injury. Seek medical personnel for guidance on how to transport.

d. Either two or four service members (head/foot) may be used to lift a litter. To lift the litter, follow the procedure below.

(1) Raise the litter at the same time as the other carriers/ bearers.

(2) Keep the casualty as level as possible.

NOTE

Use caution when transporting on a sloping incline/hill.

GLOSSARY

ACRONYMS, ABBREVIATIONS, AND DEFINITIONS

AC hydrogen cyanide
AFMAN Air Force Manual
AOC area of concentration
AR Army regulation
ATM advanced trauma management
ATNAA Antidote Treatment, Nerve Agent, Autoinjector
attn attention

BDO battle dress overgarment
BDU battle dress uniform
BZ anticholinergic drugs

C Celsius
CANA Convulsant Antidote for Nerve Agent
CASEVAC casualty evacuation
cc cubic centimeter
CG phosgene
CHS combat health support
CK cyanogen chloride
Cl chlorine
CLS Combat Lifesaver
CNS central nervous system
CO_2 carbon dioxide
COSR combat and operational stress reactions
CSR combat stress reaction
CTA common table of allowance
CX phosgene oxime

DA Department of the Army
DD Department of Defense
DM diphenylaminochloroarsine (adamsite)
DNBI disease and nonbattle injury
DOD Department of Defense
DP diphosgene
DS direct support

EMT emergency medical treatment

F Fahrenheit
FM field manual

H mustard

HD mustard
HM Hospital Corpsman
HN nitrogen mustard
HSS health service support

IPE individual protective equipment
IV intravenous

JSLIST Joint Services Light Weight Integrated Suit Technology

L lewisite
lasers laser means Light Amplification by Stimulated Emission of Radiation and sources include range finders, weapons/guidance, communication systems, and weapons simulations such as MILES [Multiple Integrated Laser Engagement System].
LBE load bearing equipment
LX lewisite and mustard

MCRP Marine Corps Reference Publication
MILES Multiple Integrated Laser Engagement System
ml milliliter
MOPP mission-oriented protective posture
MOS military occupational specialty
MTF medical treatment facility

NAPP Nerve Agent Pyridostigmine Pretreatment
NATO North Atlantic Treaty Organization
NBC nuclear, biological, and chemical
NCO noncommissioned officer
NTRP Navy Tactical Reference Publication

occlusive dressing air tight transparent dressing used to seal and cover wounds
oz ounce

PAM pamphlet
PS chloropicrin
PTSD post-traumatic stress disorder

QSTAG Quadripartite Standardization Agreement

SOP standing operating procedure
STANAG standardization agreement
STP soldier training publication

2 PAM Cl pralidoxime chloride
TB MED technical bulletin medical
TM technical manual
TSOP tactical standing operating procedure

US United States

WP white phosphorus

REFERENCES

DOCUMENTS NEEDED

These documents must be available to the intended users of this publication.

NATO STANAGs

These agreements are available on request using DD Form 1425 from Standardization Document Order Desk, 700 Robin Avenue, Building 4, Section D, Philadelphia, Pennsylvania 19111-5094.

2122. *Medical Training in First Aid, Basic Hygiene and Emergency Care.* 10 December 1975.
2126. *First Aid Kits and Emergency Medical Care Kits.* 27 September 1983.
2358. *First Aid and Hygiene Training in NBC Operations.* 3 March 1989.
2871. *First Aid Material for Chemical Injuries.* 10 March 1986.

ABCA QSTAGs

These agreements are available on request using DD Form 1425 from Standardization Document Order Desk, 700 Robin Avenue, Building 4, Section D, Philadelphia, Pennsylvania 19111-5094.

535. *Medical Training in First Aid, Basic Hygiene and Emergency Care.* 12 November 1979.

Joint and Multiservice Publications

FM 21-10. *Field Hygiene and Sanitation.* MCRP 4-11.1D. 21 June 2000.

Army Publications

AR 350-41. *Training In Units.* 19 March 1993.
DA PAM 350-59. *Army Correspondence Course Program Catalog.* 26 October 2001.
FM 3-4. *NBC Protection.* FMFM 11-9. 29 May 1992 (Reprinted with basic including Change 1, 28 October 1992; Change 2, 26 February 1996.)
FM 3-5. *NBC Decontamination.* MCWP 3-37.3. 28 July 2000. (Change 1, 31 January 2002.)

FM 3-100. *Chemical Operations Principles and Fundamentals*. MCWP 3-3.7.1. 8 May 1996.

FM 4-02.33 (8-33). *Control of Communicable Diseases Manual* (17th Edition). 3 January 2000.

FM 8-10-6 (4-02.2). *Medical Evacuation in a Theater of Operations—Tactics, Techniques, and Procedures*. 14 April 2000.

FM 4-02.7 (8-10-7). *Health Service Support in a Nuclear, Biological, and Chemical Environment*. 1 October 2002.

FM 8-284 (4-02.284). *Treatment of Biological Warfare Agent Casualties*. NAVMED P-5042; AFMAN (I) 44-156; MCRP 4-11.1C. 17 July 2000. (Change 1, 8 July 2002.)

FM 8-285 (4-02.285). *Treatment of Chemical Agent Casualties and Conventional Military Chemical Injuries*. NAVMED P-5041; AFJMAN 44-149; FMFM 11-11. 22 December 1995.

FM 22-51 (4-02.22). *Leaders' Manual for Combat Stress Control*. 29 September 1994.

CTA 8-100. *Army Medical Department Expendable/Durable Items*. 31 August 1994.

CTA 50-900. *Clothing and Individual Equipment*. 1 September 1994.

STP 21-1-SMCT. *Soldier's Manual of Common Tasks Skill Level I*. 1 October 2001.

Department of Defense Forms

DD Form 1425. *Specifications and Standards Requisition*. March 1986.

READINGS RECOMMENDED

These readings contain relevant supplemental information.

Joint and Multiservice Publications

FM 8-9 (4-02.11). *NATO Handbook on the Medical Aspects of NBC Defensive Operations AMEDP-6 (B), Part I—Nuclear, Part II—Biological, Part III—Chemical*. NAVMED P-5059; AFJMAN 44-151V1V2V3. 1 February 1996.

TB MED 81. *Cold Injury*. NAVMED P-5052-29; AFP 161-11. 30 September 1976.

TB MED 507. *Occupational and Environmental Health Prevention, Treatment and Control of Heat Injury*. NAVMED P-5052-5; AFP 160-1. 25 July 1980.

Army Publications

AR 310-25. *Dictionary of United States Army Terms (Short Title: AD)*. 15 October 1983. (Reprinted with basic including Change 1, 21 May 1986.)

AR 310-50. *Authorized Abbreviations and Brevity Codes*. 15 November 1985.

TM 3-4230-216-10. *Operator's Manual for Decontaminating Kit, Skin: M258A1 (NSN 4230-01-101-3984) and Training Aid, Skin Decontaminating: M58A1 (6910-01-101-1768)*. 17 May 1985. (Change 1, 22 January 1997.)

INDEX

References are to paragraph numbers except where specified otherwise.

Field Manual
No. 4-02.51 (8-51)

Headquarters
Department of the Army
Washington, DC, 6 July 2006

COMBAT AND OPERATIONAL STRESS CONTROL

Contents

Distribution Restriction: Approved for public release; distribution is unlimited.

***This publication supersedes FM 8-51, 29 September 1994.**

Figures

Tables

Preface

This publication outlines the functions and operations of each combat and operational stress control (COSC) element within an area of operations (AO). This field manual (FM) establishes Army doctrine and provides guidance for conducting COSC support for combat, stability, and reconstruction operations from brigade to theater level. The information provided in this publication will assist commanders and their staffs to operate efficiently at all levels of command and throughout the operational continuum. It may be used by medical planners to supplement FM 4-02, FM 8-42 (4-02.42), and FM 8-55 (4-02.55). Users of this publication should be familiar with Army Regulation (AR) 40-216 and with FM 4-02, FM 4-02.6, FM 4-02.10, FM 4-02.21, FM 4-02.24, FM 6-22.5, FM 8-10-6, FM 8-10-14, and FM 22-51.

This manual is in consonance with FM 7-15, Army Universal Task List (AUTL) and support the following Army tactical task (ART) provided below. Commanders should use the AUTL as a cross-reference for tactical tasks. The AUTL provides a standard doctrinal foundation and catalogue of the Army's tactical collective tasks.

AUTL ART	
ART 6.5.1	Provide Combat Casualty Care
ART 6.5.1.5	Provide Mental Health/Neuropsychiatric Treatment
ART 6.5.4	Provide Casualty Prevention
ART 6.5.4.5	Provide Combat Operational Stress Control Prevention

The staffing and organization structure presented in this publication reflects those established in the base tables of organization and equipment (base TOE) and are current as of the publication print date. Such staffing is subject to change to comply with manpower requirements criteria outlined in AR 71-32. Those requirements criteria are also subject to change if the modification table of organization and equipment (MTOE) is significantly altered.

Users of this publication are encouraged to submit comments and recommendations to improve the publication. Comments should include the page, paragraph, and line(s) of the text where the change is recommended. The proponent of this publication is the United States (US) Army Medical Department Center and School (USAMEDDC&S). Comments and recommendations should be forwarded in a letter format directly to the **Commander, USAMEDDC&S, ATTN: MCCS-FCD-L, 1400 East Grayson Street, Fort Sam Houston, Texas 78234-5052**, or at e-mail address: Medicaldoctrine@amedd.army.mil.

This FM applies to the Active Army, the Army National Guard (ARNG)/Army National Guard of the United States (ARNGUS), and the United States Army Reserve (USAR) unless otherwise stated.

Unless this publication states otherwise, masculine nouns and pronouns do not refer exclusively to men.

Use of trade or brand names in this publication is for illustrative purposes only and does not imply endorsement by the Department of Defense (DOD). The Army Medical Department (AMEDD) is in a transitional phase with terminology. This manual uses the most current terminology; however, other FM 4-02-series and FM 8-series may use the older terminology. Changes in terminology are a result of adopting the terminology currently used in the joint and/or North Atlantic Treaty Organization (NATO) and American, British, Canadian, and Australian (ABCA) Armies publication arenas. Therefore, the following terms are synonymous and the current terms are listed first, to include—

- Force health protection (FHP), health service support (HSS), and combat health support (CHS).
- Medical logistics (MEDLOG), health service logistics (HSL), and combat health logistics (CHL).
- Levels of care, echelons of care, and roles of care.

- Combat and operational stress control (COSC) and combat stress control (CSC).
- Behavioral health (BH) and mental health (MH).

Additionally, please note that the term "battle fatigue (BF)" that is used in AR 40-216 is being replaced with the term "combat stress reaction (CSR)." This change is due to DOD Directive (DODD) 6490.5, which specified that all Military Services use the term CSR for the purpose of joint interoperability. A DOD (Health Affairs) working group with the Services later added the term "operational stress reaction (OSR)" to further characterize stress reactions experienced by Soldiers. The only difference between a CSR and an OSR is that it takes place in a combat environment. To reduce confusion from this change in terminology, this manual will use combat and operational stress reaction (COSR) as opposed to "BF". The COSR (battle fatigue) "casualties" are Soldiers experiencing a stress reaction in combat or operational environment.

Introduction

In our own Soldiers and in the enemy combatants, control of stress is often the decisive difference between victory and defeat across the operational continuum. Battles and wars are won more by controlling the will to fight than by killing all of the enemy combatants. Uncontrolled combat stress causes erratic or harmful behaviors, impair mission performance, and may result in disaster and defeat of COSC preventive measures. The COSC preventive measures are aimed at minimizing maladaptive stress reactions while promoting adaptive stress reactions, such as loyalty, selflessness, and acts of bravery.

This manual provides doctrinal guidance for controlling excessive stress in combat and other operational environments. It identifies command and leadership responsibilities for COSC. It identifies COSC consultation, training, and education assistance available for units. This manual provides definitive guidance to BH personnel and CSC units for their COSC mission and for management of COSR and other behavioral disordered patients (BDPs). It identifies the requirements for COSC consultation, planning, coordination, rehearsal, and implementation of the COSC plan contained in the FHP annex of the operation order (OPORD).

Many stressors in a combat situation are due to deliberate enemy actions aimed at killing, wounding, or demoralizing our Soldiers and our allies. Other stressors are due to the operational environment. Some of these stressors can be avoided or counteracted by wise command actions. Still other stressors are due to our own calculated or miscalculated choice, accepted in order to exert greater stress on the enemy. Sound leadership works to keep stressors within tolerable limits and prepares the troops mentally and physically to endure them. Some of the most potent stressors can be due to personal organizational problems in the unit or on the home front. These, too, must be identified and, when possible, corrected or controlled. Unit needs assessments (UNAs) can help BH providers identify specific stressor in a unit and develop interventions to help unit personnel cope.

This manual identifies MH sections and medical units, CSC, that provides COSC support to units. It identifies the requirements for COSC assistance, as a unit transitions through the different phases of a deployment. Chaplains, commander, leader and COSC Soldier mentors provide assistance with after-action review (AAR) and traumatic events management (TEM).

Chapter 1
Combat and Operational Stress Control

SECTION I — INTERVENTION AND CONTROL FOR THE COMBAT AND OPERATIONAL STRESS THREAT

STRESS CONTROL

1-1. Combat stress includes all the physiological and emotional stresses encountered as a direct result of the dangers and mission demands of combat, see AR 40-216. Combat and operational stress control in the Army may be defined as programs developed and actions taken by military leadership to prevent, identify, and manage adverse COSRs in units. This program optimizes mission performance; conserves the fighting strength; and prevents or minimizes adverse effects of COSR on Soldiers and their physical, psychological, intellectual, and social health. Its goal is to return Soldiers to duty expeditiously. According to DODD 6490.2, COSC activities include routine screening of individuals when recruited; continued surveillance throughout military service, especially before, during, and after deployment; continual assessment and consultation with medical and other personnel from garrison to the battlefield.

1-2. Combat and operational stress control is the commander's responsibility at all levels. The commander is assisted with his responsibility for COSC by his staff, unit leaders, unit chaplain (Appendix A), and organic medical personnel. The commander may also receive assistance from organic COSC personnel at brigade and above, and from corps and above medical company/detachment CSC BH personnel. The key concern to combat commanders is to maximize the return-to-duty (RTD) rate of Soldiers who are temporarily impaired or incapacitated with stress-related conditions or diagnosed behavioral disorders.

1-3. The purpose of COSC is to promote Soldier and unit readiness by—

- Enhancing adaptive stress reactions.
- Preventing maladaptive stress reactions.
- Assisting Soldiers with controlling COSRs.
- Assisting Soldiers with behavioral disorders.

Note. The word control is used with combat and operational stress, rather than the word management, to emphasize the active steps that leaders, supporting BH personnel, and individual Soldiers must take to keep stress within an acceptable range.

COMBAT AND OPERATIONAL STRESS THREAT

1-4. Many stressors in a combat situation are due to deliberate enemy actions aimed at killing, wounding, or demoralizing our Soldiers and our allies. Other stressors are due to the natural environment, such as intense heat or cold, humidity, or poor air quality. Still others are due to leaders' own calculated or miscalculated choices (for example, decisions about unit strength, maneuver, the time of the attack, and plans for medical and logistical support). Sound leadership works to keep operational stressors within tolerable limits and prepares troops mentally and physically to endure them. In some cases however, excessive stress can affect both leaders' and Soldiers' decision-making and judgment, resulting in missed opportunities, or worse, in high casualties and/or failure to complete the mission. Finally, some of the most potent stressors are interpersonal in nature and can be due to conflict in the unit or on the home front. Extreme reactions to such stressors may involve harm to self (as in the hypothetical case of a Soldier that becomes suicidal on discovering that his wife wants a divorce) or to others (as in the case of a Soldier that impulsively fires a weapon at his noncommissioned officer (NCO) out of rage over perceived unfairness). These stressors must be identified and when possible, corrected or controlled. For more information on the control of combat/operational stressors and for details about specific leader actions to control stress, see FM 22-51. Also see FM 4-02 and FM 4-02.17 for additional information on the general and medical threat.

EFFECT OF STRESS

1-5. Focused stress is vital to survival and mission accomplishment. However, stress that is too intense, or prolonged, results in COSR that impairs their ability to function effectively. Some stressors contribute to misconduct that requires disciplinary action and may take the Soldier from duty for legal action and incarceration. In a broader context stress may cause battle and nonbattle injuries through inattention, clumsiness, and reckless behavior. These resultant injuries can include equipment losses and friendly fire incidents. Stress may increase disease rates by disrupting hygiene and protective measures, and impairing the body's immune defenses. Stress may progress to BH disorders, or suicidal behaviors and/or homicidal behaviors. Excessive stress in combat contributes to lapses in operational and tactical judgment and to missed opportunities that could increase the numbers of Soldiers injured over time.

MODEL FOR COMBAT AND OPERATIONAL STRESS CONTROL INTERVENTIONS

1-6. Combat and operational stress control assessments are performed during all phases of combat operations, stability and reconstruction operations, and support operations so the term COSC may refer to any of these different types of Army operations. The COSC assessments are performed at unit and individual level. They consider the range of variables according to a model, which recognizes that biological, psychological, and social factors influence each other. The COSC assessment reviews the interaction systematically to a depth appropriate to the need. The assessment identifies which variables can be modified to improve coping or outcome. Based on these assessments COSC personnel recommend courses of action (COA) to the commander. They identify and initiate COSC interventions to improve unit effectiveness and Soldier efficiency and well-being.

1-7. Provided in Figure 1-1 is a conceptual model of stress, its mitigating and aggravating factors and potential outcomes on Soldiers and families. This model can be helpful when designing COSC interventions to improve short-term and long-term outcomes.

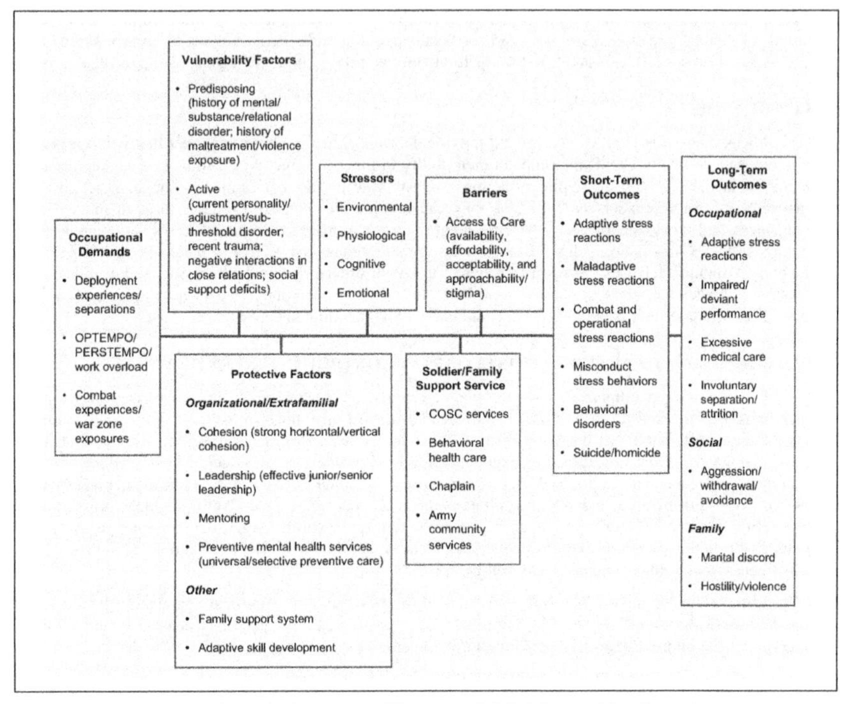

Figure 1-1. Model of stress and its potential Soldier and family outcomes

MENTAL AND PHYSICAL STRESSORS

1-8. A rough distinction between a mental and physical stressor can be made—

- A mental stressor is one in which the brain receives information about a given threat or demand, but this information results in only indirect physical impact on the body. Instead, its primary effect is to place demands on and evoke reactions from the perceptual, cognitive and/or emotional systems of the brain (such as information overload, perceived lack of control, or grief-producing losses).
- A physical stressor has a direct, potentially harmful effect on the body. These stressors may be external environmental conditions (such as temperature) or the internal physiologic demands required by or placed upon the human body (such as the need for hydration, or an immune response to a viral infection).

TYPES OF MENTAL AND PHYSICAL STRESSORS

1-9. Figure 1-2 provides examples for the two types of physical stressors (environmental and physiological) and the two types of mental stressors (cognitive and emotional). Also, as recognized in the COSC intervention model (Figure 1-1) physical stressors cause mental stressors when they result in discomfort, distraction, and threat of harm, as well as when they directly impair brain functions. Mental stressors can lead to adaptive or maladaptive stress behaviors that decrease or increase the exposure to physical stressors.

PHYSICAL STRESSORS	MENTAL STRESSORS
ENVIRONMENTAL	**COGNITIVE**
Heat, cold, wetness, dust Vibration, noise, blast Noxious odors (fumes, poisons, chemicals) Directed-energy weapons/devices Ionizing radiation Infectious agents Physical work Poor visibility (bright lights, darkness, haze) Difficult or arduous terrain High altitude	Information (too much or too little) Sensory overload or deprivation Ambiguity, uncertainty, unpredictability Time pressure or waiting Difficult decision (rules of engagement) Organizational dynamics and changes Hard choices versus no choice Recognition of impaired functioning Working beyond skill level Previous failures
PHYSIOLOGICAL	**EMOTIONAL**
Sleep deprivation Dehydration Malnutrition Poor hygiene Muscular and aerobic fatigue Overuse or underuse of muscles Impaired immune system Illness or injury Sexual frustration Substance use (smoking, caffeine, alcohol) Obesity Poor physical condition	Being new in unit, isolated, lonely Fear and anxiety-producing threats (of death, injury, failure, or loss) Grief-producing losses (bereavement) Resentment, anger, and rage-producing frustration and guilt Inactivity producing boredom Conflicting/divided motives and loyalties Spiritual confrontation or temptation causing loss of faith Interpersonal conflict (unit, buddy) Home-front worries, homesickness Loss of privacy Victimization/harassment Exposure to combat/dead bodies Having to kill

Figure 1-2. Examples of combat and operational stressors

STRESS BEHAVIORS IN COMBAT AND OTHER OPERATIONS

1-10. Combat and operational stress behavior is the generic term that is used for the full spectrum of combat and operational stress behaviors. It covers the range of reactions, from adaptive to maladaptive. Figure 1-3 provides a listing of typical adaptive and maladaptive stress reactions.

ADAPTIVE STRESS REACTIONS

1-11. Stressors, when combines with effective leadership and good peer relationships may lead to adaptive stress reactions which enhance individual and unit performance. Examples of adaptive stress reactions include—

- The strong personal trust, loyalty, and cohesiveness (called horizontal bonding), which develops among peers in a small military unit.
- Personal trust, loyalty, and cohesiveness that develops between leaders and subordinates (called vertical bonding).
- Esprit de corps is also defined as a feeling of identification and membership in the larger, enduring unit with its history and ideas. This may include the unit (such as battalion, brigade

combat team [BCT], regiment, or division) the branch (such as infantry, artillery, or military police [MP]) and beyond the branch to the US Army level.

- Unit cohesion is the binding force that keeps Soldiers together and performing the mission in spite of danger and death. Cohesion is a result of Soldiers knowing and trusting their peers and leaders and understanding their dependency on one another. It is achieved through personal bonding and a strong sense of responsibility toward the unit and its members. The ultimate adaptive stress reactions are acts of extreme courage and almost unbelievable strength. They may even involve deliberate self-sacrifice.

MALADAPTIVE STRESS REACTIONS

1-12. Combat and operational stress reaction and misconduct stress behaviors comprise the maladaptive stress reactions.

COMBAT AND OPERATIONAL STRESS REACTION

1-13. The Army uses the DOD-approved term/acronym COSR in official medical reports. This term can be applied to any stress reaction in the military unit environment. Many reactions look like symptoms of mental illness (such as panic, extreme anxiety, depression, hallucinations), but they are only transient reactions to the traumatic stress of combat and the cumulative stresses of military operations. Some individuals may have behavioral disorders that existed prior to deployment or disorders that were first present during deployment, and need BH intervention beyond the interventions for COSR.

MISCONDUCT STRESS BEHAVIORS

1-14. Examples of misconduct stress behaviors are listed in Figure 1-3. These range from minor breaches of unit orders or regulations to serious violations of the Uniform Code of Military Justice (UCMJ) and the Law of Land Warfare. Misconduct stress behaviors are most likely to occur in poorly trained undisciplined Soldiers, however good and heroic, under extreme combat stress may also engage in misconduct. Generally, misconduct stress behaviors—

- Range from minor breaches of unit orders or regulations to serious violations of the UCMJ and the Law of Land Warfare.
- May also become a major problem for highly cohesive and proud units. Such units may come to consider themselves entitled to special privileges and, as a result, some members may relieve tension unlawfully when they stand-down from their military operations. They may lapse into illegal revenge when a unit member is lost in combat. Stress control measures and sound leadership can prevent such misconduct stress behaviors, but once serious misconduct has occurred, Soldiers must be punished to prevent further erosion of discipline. Combat stress, even with heroic combat performance, cannot justify criminal misconduct.

OVERLAPPING STRESS BEHAVIORS

1-15. The distinctions among adaptive stress reactions, misconduct stress behaviors, and COSRs are not always clear. Indeed, the three categories of combat and operational stress behaviors may overlap. Soldiers with COSR may show misconduct stress behaviors and vice versa. Soldiers with adaptive stress reactions may also suffer from COSR. Finally, excellent combat Soldiers that have exhibited bravery and acts of heroism may also commit misconduct stress behaviors.

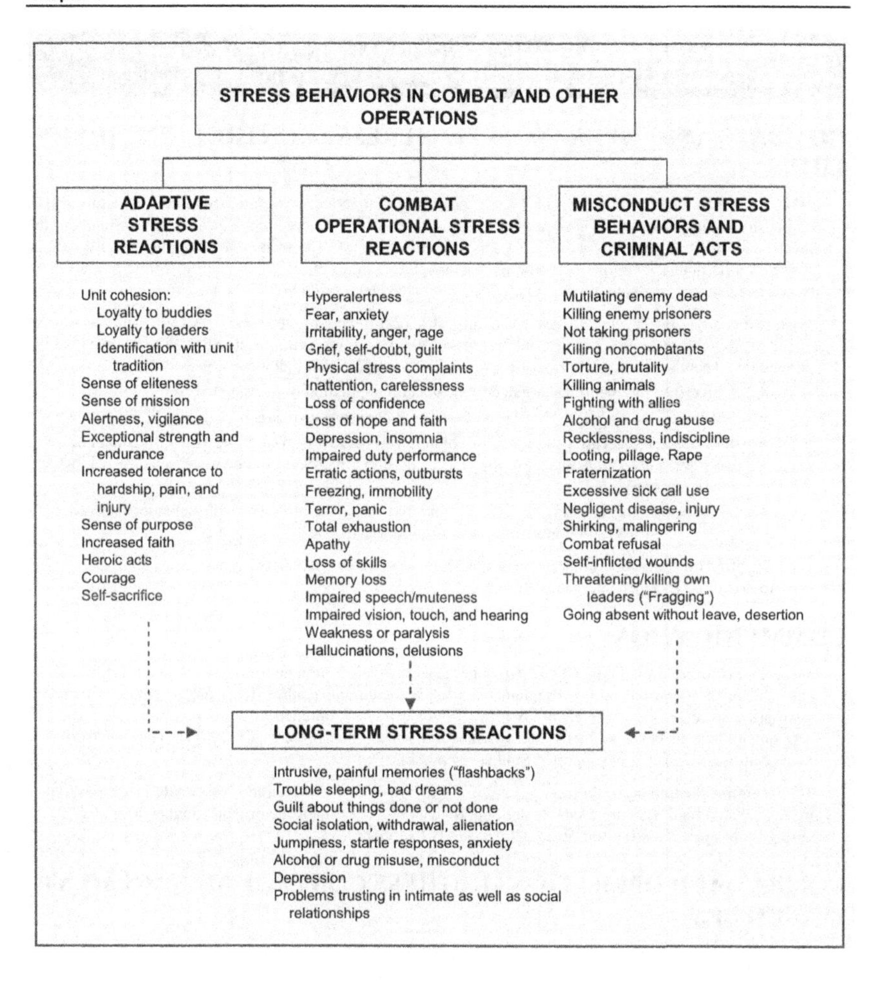

Figure 1-3. Stress behaviors in combat and other operations

1-16. It is common for stress reactions to persist or arise long after exposure to distressing events. When there is impairment in social and/or occupational functioning, a clinical assessment is warranted. Combat and operational stress control is important to sustain Army strength over the long-term and reduce the cost to society, the DOD, the Soldiers, and families.

SECTION II — PRINCIPLES AND FUNCTIONAL AREAS FOR COMBAT AND OPERATIONAL STRESS CONTROL

COMBAT AND OPERATIONAL STRESS CONTROL INTERVENTIONS

1-17. Soldier and unit readiness is best achieved through an active, prevention-focused orientation that is embodied in the principles. These principles apply to all COSC interventions or activities throughout the theater and are followed by medical personnel in all BH/COSC elements. Their application may differ based on a particular level of care and other factors pertaining to the mission, enemy, terrain and weather, troops and support available, time available, and civil considerations (METT-TC).

1-18. Preventive interventions seek to reduce the occurrence or severity of COSR and behavioral disorders, thereby sustaining Soldier and unit readiness. These interventions are tailored to the needs of the population. There are four categories of preventive interventions that include—

- Universal. Interventions targeted to the general population or an assigned AO.
- Selective. Interventions targeted to a unit or Soldier whose risk is higher than average.
- Indicated. Interventions targeted to Soldiers with COSR or indications of a potential behavioral disorder, and to units that show signs their mission effectiveness is being affected by combat and operational stressors.
- Treatment. Interventions targeted to treat and follow-up Soldiers with behavioral disorders to prevent their loss from duty.

1-19. Identify life- or function-threatening medical, surgical, or psychiatric condition as soon as possible and provide those patients emergency treatment.

COMMUNICATIONS

1-20. To maximize prevention, COSC personnel must maintain a high degree of involvement with Soldiers and leaders of supported units. Regular visits to the battalion/squadron level are essential to maintain appropriate accessibility. The COSC personnel must always be attuned to Soldiers around them and not rely on Soldiers to come to them. This means that when time permits, COSC personnel need to start conversations with as many supported Soldiers as possible.

1-21. Communication is the delivery system of COSC services and therefore is essential. Use all available means to communicate/coordinate with supported units, subordinate teams, and higher headquarters to ensure delivery of effective and timely COSC and BH services.

COMBAT AND OPERATIONAL STRESS CONTROL MANAGEMENT PRINCIPLES

1-22. The COSC management principles of brevity, immediacy, contact, expectancy, proximity, and simplicity (BICEPS). These principles apply to all COSC interventions or activities throughout the theater, and are followed by COSC personnel in all BH/COSC elements. These principles may be applied differently based on a particular level of care and other factors pertaining to METT-TC.

Note. The BICEPS principles are discussed in-depth in FM 6-22.5. This is a multiservice publication with the US Marine Corps (USMC). Due to differences in organizational structure the USMC defines the "C" in BICEPS to mean *centrality*, while the US Army defines it to mean *contact*, which will be used for the remainder of this manual.

Brevity, Immediacy, Contact, Expectancy, Proximity, and Simplicity

1-23. Using BICEPS is extremely important in the management of Soldiers with COSR and/or behavioral disorders.

Brevity

1-24. Initial rest and replenishment at COSC facilities located close to the Soldier's unit should last no more than 1 to 3 days (USMC and Navy is 3 to 4 days). Those requiring further treatment are moved to the next level of care. Since many require no further treatment, military commanders expect their Soldiers to RTD rapidly.

Immediacy

1-25. It is essential that COSC measures be initiated as soon as possible when operations permit. Intervention is provided as soon as symptoms appear.

Contact

1-26. The Soldier must be encouraged to continue to think of himself as a warfighter, rather than a *patient* or a *sick person*. The chain of command remains directly involved in the Soldier's recovery and RTD. The COSC team coordinates with the unit's leaders to learn whether the overstressed individual was a good performer prior to the COSR. Whenever possible, representatives of the unit or messages from the unit tell the Soldier that he is needed and wanted back. The COSC team coordinates with the unit leaders, through unit medical personnel or chaplains, any special advice on how to assure quick reintegration when the Soldier returns to his unit.

Expectancy

1-27. The individual is explicitly told that he is reacting normally to extreme stress and is expected to recover and return to full duty in a few hours or days. A military leader is extremely effective in this area of treatment. Of all the things said to a Soldier suffering from COSR the words of his small-unit leader have the greatest impact due to the positive bonding process that occurs. A simple statement from the small-unit leader to the Soldier that he is reacting normally to COSR and is expected back soon have positive impact. Small-unit leaders should tell Soldiers that their comrades need and expect them to return. When they do return, the unit treats them as every other Soldier and expects them to perform well.

Proximity

1-28. Soldiers requiring observation or care beyond the unit level are evacuated to facilities in close proximity to, but separate from the medical or surgical patients at the battalion aid station (BAS) or medical company nearest the Soldiers' unit. It is best to send Soldiers who cannot continue their mission and require more extensive intervention to a facility other than a hospital, unless no other alternative is possible. Combat and operational stress reactions are often more effectively managed in areas close to the Soldier's parent unit. On the noncontiguous battlefield characterized by rapid, frequent maneuver and continuous operations, COSC personnel must be innovative and flexible in designing interventions which maximize and maintain the Soldier's connection to his parent unit. See FM 6-22.5 for additional definition of proximity.

Simplicity

1-29. Indicates the need to use brief and straightforward methods to restore physical well-being and self-confidence.

1-30. The actions used for COSR control (commonly referred to as the 5 R's) involve the following actions:

- **R**eassure of normality.
- **R**est (respite from combat or break from the work).

- Replenish bodily needs (such as thermal comfort, water, food, hygiene, and sleep).
- Restore confidence with purposeful activities and contact with his unit.
- Return to duty and reunite Soldier with his unit.

Note. Historically the Army had used the terms proximity, immediacy, expectancy and simplicity (PIES) but began using BICEPS when it became the approved joint terminology.

NONPATIENT STATUS

1-31. To prevent Soldiers with COSR from adopting the *patient* role, these guidelines should be followed:

- Keep the Soldier in uniform and hold him responsible for maintaining Soldier standards.
- Keep the Soldier separate from seriously ill or injured patients.
- Avoid giving him medications unless essential to manage sleep.
- Do not evacuate or hospitalize the Soldier unless absolutely necessary.
- Do not diagnose the Soldier prematurely.
- Transport the Soldier via general-purpose vehicles, not ambulances.

COMBAT AND OPERATIONAL STRESS CONTROL FUNCTIONAL AREAS

1-32. Combat and operational stress control interventions and activities are organized into nine functional areas. These functional areas cover the full spectrum of BH care from preventive through clinical intervention. They are defined below and are discussed further in subsequent chapters of this manual.

UNIT NEEDS ASSESSMENT

1-33. Unit needs assessment is the systematic and frequent assessment of supported units to determine the priority and types of BH interventions required. See Chapter 4.

CONSULTATION AND EDUCATION

1-34. Consultation involves the liaison with and preventive advice to commanders, staff of supported units, and Soldiers. Education involves training in concepts and skills for increasing Soldier resilience to stress. See Chapter 5.

TRAUMATIC EVENTS MANAGEMENT

1-35. Traumatic events management blends other COSC functional areas to create a flexible set of interventions specifically focused on stress management for units and Soldiers following potentially traumatizing events (PTE). Like other functional areas, COSC providers must tailor TEM to the needs of the unit and the Soldier. See Chapter 6.

RECONSTITUTION SUPPORT

1-36. Reconstitution is extraordinary action that commanders plan and implement to restore units to a desired level of combat effectiveness commensurate with mission requirements and available resources. It transcends normal day-to-day force sustainment actions. However, it uses existing systems and units to do so. No resources exist solely to perform reconstitution. In COSC reconstitution support, COSC personnel are responsible for providing unit Soldier restoration and conducting the COSC functions. This support is provided to units following traumatic events and during reconstitution, redeployment, and transition among levels of operational tempo (OPTEMPO). See Chapter 7.

Combat and Operational Stress Control Triage

1-37. Combat and operational stress control triage is the process of sorting Soldiers with COSR and/or BH disorders based upon where they can best be managed. See Chapter 8.

Combat and Operational Stress Control Stabilization

1-38. This function provides initial management of Soldiers with severe COSR or behavioral disorders. Their safety is ensured and they are evaluated for RTD potential or prepared for further treatment or evacuation. See Chapter 9.

Soldier Restoration

1-39. Soldier restoration involves the one- to three-day management of Soldiers with COSR or behavioral disorders normally near a medical treatment facility (MTF) in close proximity to his unit. This approach uses the 5 R's discussed above. See Chapter 10.

Behavioral Health Treatment

1-40. Patients with identified behavioral disorders receive ongoing evaluation, treatment and follow-up to sustain them. This functional area implies a therapist-patient relationship, clinical documentation, and adherence to clinical standards of care. See Chapter 11.

Soldier Reconditioning

1-41. Reconditioning is an intensive program of work therapy, military activities, physical training and psychotherapy. Reconditioning programs are conducted up to seven days (or more) in the corps area. Additional reconditioning may be provided in the theater outside the combat zone (CZ). See Chapter 12.

Priority of Functional Areas

1-42. The medical company/detachment, CSC commanders, MH section leaders, senior medical commanders, and command surgeons must set priorities, coordinate their actions, and allocate resources to accomplish missions based upon the total situation/METT-TC. Needs assessments and available resources guide COSC interventions. At the unit level, the unit needs assessment helps determine which functional areas (such as consultation and education; and critical event and transition management) take priority to achieve success with the COSC mission. At the Soldier level, COSC triage helps determine which functional areas (such as consultation and education, stabilization, Soldier restoration, behavioral health treatment, and reconditioning) take priority to achieve COSC mission goals.

Chapter 2

Behavioral Health and Combat and Operational Stress Control Elements in the Theater

SECTION I — MENTAL HEALTH SECTIONS

ASSIGNMENTS

2-1. Mental health sections are located in medical companies assigned to brigades, divisions, corps, and theater-level medical units.

ARMY OF EXCELLENCE

2-2. In the Army of Excellence (AOE) divisions, the division MH section (DMHS) is organic to the main support medical company (MSMC). In the armored cavalry regiments (ACRs) and separate brigades, a MH section is assigned to the forward support medical company (FSMC) of the forward support battalion (FSB) or a medical troop. In the corps and echelons above corps (EAC), the MH sections are located in the headquarters and headquarters detachment (HHD), area support medical battalion (ASMB) and in the area support medical companies (ASMCs). In those AOE divisions that are undergoing transformation, the MH sections are assigned to each of the brigade support medical companies (BSMCs) of the brigade support battalion (BSB) and in the ASMCs from the corps ASMB.

MODULAR FORCE

2-3. In the modular BCTs, Stryker BCTs (SBCTs) and the heavy BCTs (HBCTs) MH sections are assigned to the BSMC of the BSB. At echelons above brigade (EAB) and at the division, corps, or theater levels, the MH sections are assigned to the multifunctional medical battalion (MMB).

> ***Note.*** Under the modular design, the MMB will replace the ASMB, the medical logistics battalion, and the medical evacuation battalion.

ORGANIZATIONS

2-4. The AOE divisions have only one MH section that is assigned to the MSMC of the main support battalion (MSB), which is located in the division support area (DSA). See Table 2-1 for a listing of personnel assigned to the division MH section. These personnel are deployed to the supported brigade as two-man teams comprised of one officer and one MH specialist.

Table 2-1. Division mental health staff

Division mental health staff
Division Psychiatrist (MAJ, Area of Concentration [AOC] 60W00, Medical Corps [MC])
Social Work Officer (CPT, AOC 73A67, Medical Service [MS] Corps)
Clinical Psychologist (CPT, AOC 73B67, MS)
Mental Health NCO (E7, E6, and E5, Military Occupational Specialty [MOS] 68X40/30/20)
Mental Health Specialist (Four)—(Two E4 and Two E3, MOS 68X10)

FUNCTIONS AND RESPONSIBILITIES OF ALL MENTAL HEALTH SECTIONS

2-5. All MH sections regardless of their organizational assignment are tasked with providing COSC for their supported units. In all of these units, COSC is accomplished through vigorous prevention, consultation, training, educational, and Soldier restoration programs. These programs are designed to provide BH expertise to unit leaders and Soldiers where they serve and sustain their mission focus and effectiveness under heavy and prolonged stress. The MH sections identify Soldiers with COSRs who need to be provided rest/Soldier restoration within or near their unit area for rapid RTD. These programs are designed to maximize the RTD rate of Soldiers who are either temporary impaired, have a diagnosed behavioral disorder, or have stress-related conditions. Also the prevention of posttraumatic stress disorders (PTSDs) is an important objective for brigades and EAB. The behavioral health officer (either a clinical psychologist or social work officer) and MH specialist are especially concerned with assisting and training of—

- Small-unit leaders.
- Unit ministry teams (UMTs) and staff chaplains.
- Battalion medical platoons.
- Patient-holding squad and treatment squad personnel of the medical company.

RESPONSIBILITIES

2-6. The MH section has a primary responsibility for assisting commanders with COSC by implementing the brigade combat mental fitness program. Also the MH section serves as a consultant to the commander, staff, and others involved with providing prevention and intervention services to unit Soldiers and their families. The MH section has staff responsibilities for assisting the brigade surgeon with establishing brigade policy and guidance for the prevention, diagnosis, treatment, management, and RTD of stress-related casualties. This is accomplished under the guidance and in close coordination with maneuver battalions and FSMC/BSMC physicians.

FUNCTIONS

2-7. The COSC functional areas were identified in Chapter 1. The functions of all MH sections are to support the nine functional areas except Soldier reconditioning.

UTILIZATION IN GARRISON

2-8. In garrison, BH personnel assigned to the division, brigade, or corps/EAC units continue to perform the same staff and outreach functions with supported units as they do in a field environment. An increase in the BH treatment functions may be possible as a result of consolidating BH care providers. The BH providers make available their consultation skills and clinical expertise to the Soldiers of supported units and their family readiness groups. Clinical care of family members and of Soldiers that require longer-term care beyond crisis intervention, brief treatment, and medication follow-up is the responsibility of the medical department activity (MEDDAC)/medical center (MEDCEN). The MH section personnel should focus their clinical work primarily on Soldiers with problems amenable to brief treatment.

2-9. Clinical services may be provided as part of a consolidated BH activity (as is usually established by a DMHS) or by augmenting an existing MEDDAC/MEDCEN BH staff.

2-10. The MH sections should strive to reduce referrals for BH treatment by working closely with unit leaders and chaplains to control organizational stress and rapidly identify and intervene with those Soldiers having BH disorders.

2-11. When the medical company or battalion deploys on training exercises, assigned BH personnel deploy with them to provide COSC training and support. In addition, they train to improve their own technical and tactical skills.

Note. In accordance with AR 40-216, clinical responsibilities in garrison must not interfere with participation in field and deployment exercises and maintenance of combat readiness.

MEDICAL DETACHMENT, COMBAT STRESS CONTROL

MISSION

2-12. The mission of the medical detachment, CSC (TOE 08463A000) is to provide COSC interventions and stress prevention activities to supported units in its AO. The staff augments division and brigade MH sections; provides direct support (DS) to combat brigades without organic BH officers; and provides area support in its AO. The medical detachment, CSC reconstitutes other brigade and division COSC assets. The medical detachment, CSC provides COSC interventions and activities to indigenous populations as directed in stability and reconstruction operations, humanitarian assistance, disaster relief, peace support operations, and detention facility operations. The medical detachment, CSC provides COSC interventions and activities to units in support of their readiness preparation and throughout their deployment cycle.

ASSIGNMENT

2-13. The Medical Reengineering Initiative (MRI) medical detachment, CSC (TOE 08463A000) is assigned to a corps medical brigade (MEDBDE) (TOE 08422A100), or other medical command and control (C2) element. Its teams may be further attached to a medical company, CSC (TOE 08467L000), an ASMB (TOE 08456A000), a BSMC/MSMC, or a combat support hospital (CSH). See Appendix B for the proposed future medical detachment, COSC.

EMPLOYMENT AND CAPABILITIES

2-14. The detachment headquarters is usually located in the DSA. Its teams disperse throughout the division's AO, and may extend support to corps if no CSC medical company is available. At TOE Level 1, the MRI medical detachment, CSC provides—

- A preventive section capable of dividing into four preventive teams. Each team provides mobile COSC interventions and activities.
- A fitness section capable of dividing into two fitness teams. Each fitness team provides mobile COSC interventions and activities. Each team is equipped to hold 40 Soldiers at the same time. With additional logistical support, each team can accommodate additional Soldiers under surge conditions.

STAFF RESPONSIBILITIES

2-15. The medical detachment, CSC assists the C2 headquarters (of the unit to which it is assigned or attached) regarding planning and coordination of COSC support, stress threat, mental and physical stressors, stress behaviors, principles of COSC, and implementation of COSC functional areas.

DEPENDENCY

2-16. This unit is dependent on—

- Appropriate elements of the corps for FHP, religious support, legal, finance, field feeding, personnel and administrative services support; laundry and clothing exchange; mortuary affairs support; and security of enemy prisoners of war (EPW), detainee, and US prisoner patients.
- The medical headquarters to which it is assigned/attached for FHP; medical administration, logistics, including MEDLOG; medical regulating of patients; evacuation; coordination for RTD; and unit level equipment and communications-electronics (CE) maintenance.

MOBILITY AND SECURITY

2-17. This unit is 100-percent mobile. It requires 100 percent of its organic personnel and equipment be transported in a single lift, using its organic vehicles. Upon relocation, Soldiers being held will require additional transportation. This unit is responsible for perimeter defense of its immediate operational area. However, it is dependent on appropriate elements of the corps for additional security, to include security of convoy operations. Personnel of the detachment are provided weapons for their personal defense and for the defense of their patients and/or held Soldiers.

ORGANIZATION

2-18. This 43-person MRI medical detachment (Figure 2-1) is organized into a headquarters section; a preventive section composed of four CSC preventive (CSCP) teams; and a fitness section consisting of two CSC fitness (CSCF) teams.

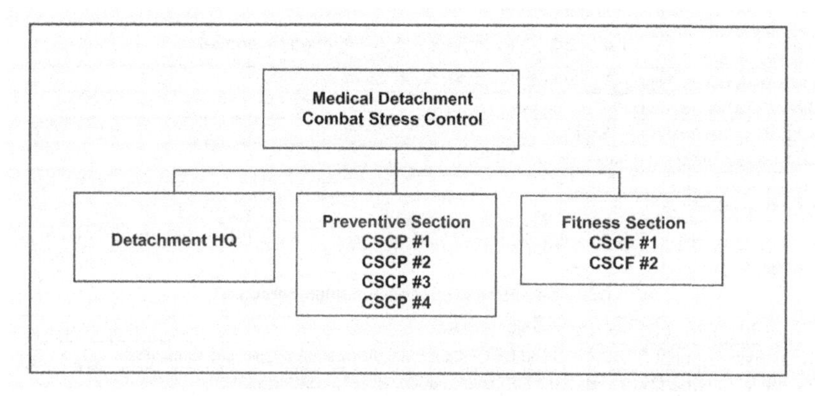

Figure 2-1. Medical Reengineering Initiative medical detachment, combat stress control (TOE 08463A000)

DETACHMENT HEADQUARTERS

2-19. The medical detachment, CSC headquarters provides C2 for the detachment. The headquarters element is responsible for planning, coordinating, and implementing COSC support for supported units. Personnel of the headquarters element provide maintenance, supply and service, and personnel administrative support. See Table 2-2 for a list of personnel assigned to the detachment headquarters.

Table 2-2. Detachment headquarters assigned personnel

Detachment Commander (LTC, AOC 60W00, MC)
Field Medical Assistant (1LT, AOC 70B67, MS)
Detachment Sergeant (E7, MOS 68X40)
Supply Sergeant (E5, MOS 92Y20)
Human Resource Specialist (E4, MOS 42A10)
Wheeled Vehicle Mechanic (E4, MOS 63B10)
Cook (E3, MOS 92G10)
Note. Detachment officers and NCOs from the preventive section and the fitness section may be assigned additional duties, which enhance the overall effectiveness of the headquarters section.

Preventive Section

2-20. The medical detachment, CSC has a 16-person preventive section. See Table 2-3 for listing of personnel assigned to the preventive section. The section can be divided into four separate CSCP teams. Three CSCP teams are normally allocated to supported maneuver brigades (one team per brigade). The fourth CSCP team may provide direct support to a SBCT, an aviation brigade, other brigade-sized units, or to corps units operating in a division's AO. Preventive section personnel may be task-organized with personnel of the fitness section into teams for specific missions. The section (and team) leader position may be held by any of the officers assigned to the section. The section's COSC interventions and activities are—

- Unit needs assessment; consultation and education; TEM; COSC triage; stabilization (emergency); and BH treatment.
- Assisting with Soldier restoration and reconditioning at the CSC detachment program. Overseeing a one- to three-day COSC Soldier restoration program in a brigade, division, or ASMC holding section or in another area suitable for Soldiers experiencing COSR and/or other stress-related disorders.

Table 2-3. Preventive section assigned personnel

Social Work Officer (Four)—(Two MAJ and Two CPT, AOC 73A67, MS)
Clinical Psychologist (Four)—(Two MAJ and Two CPT, AOC 73B67, MS)
Mental Health NCO (Four)—(E5, MOS 68X20)
Mental Health Specialist (Four)—(E4, MOS 68X10)

Fitness Section

2-21. The medical detachment, CSC has a 20-person fitness section. Personnel assigned to the fitness section are identified in Table 2-4.

Table 2-4. Fitness section assigned personnel

Psychiatrist (Two)—(LTC and MAJ, AOC 60W00, MC)
Occupational Therapist (Two)—(MAJ and CPT, AOC 65A00, Army Medical Specialty Corps [SP])
Psychiatric/BH Nurse (Two)—(MAJ and CPT, AOC 66C00, Army Nurse Corps [AN])
Occupational Therapy NCO (Four)—(Two E6 and Two E5, MOS 68W30/20N3)
Mental Health NCO (Four)—(Two E6 and Two E5, MOS 68X30/20)
Mental Health Specialist (Six)—(Four E4 and Two E3, MOS 68X10)

2-22. The fitness section is task-organized to provide COSC interventions and activities. Fitness section personnel may be task-organized with personnel of the preventive section into teams for specific missions. This section can divide into two CSCF teams. Each fitness team can deploy a four-person mobile team using the team's high-mobility multipurpose wheeled vehicle (HMMWV). One CSCF team usually collocates with a supported divisional medical company to provide mobile COSC support within a DSA and conduct Soldier restoration programs, as required. The second CSCF team augments area support to corps units in the division AO and in forward areas of the corps. The CSCF team provides staff and equipment for operating a Soldier restoration or reconditioning center. This section's COSC interventions and activities are—

- Unit needs assessment; consultation and education; TEM; COSC triage; stabilization; and BH treatment.
- Conducting Soldier restoration and reconditioning programs.
- Staffing and operating a psychiatric ward for supported CSH when this capability is required (refer to FM 8-10-14).

MEDICAL COMPANY, COMBAT STRESS CONTROL

2-23. The MRI medical company, CSC is employed in the corps and EAC. The basis of allocation is one medical company, CSC per corps or theater. The medical company, CSC is task-organized, METT-TC-dependent for stability and reconstructions operations. Medical company, CSC (TOE 08467A000) replaces medical company, CSC (TOE 08467L000). A medical company, CSC provides comprehensive preventive and treatment services to a corps and EAC during war. It provides this support to all services on an area support basis. The medical company, CSC provides DS to combat and combat support (CS) brigades, as needed. It reinforces or reconstitutes other CSC assets in the corps or divisions if required. The medical company, CSC provides COSC/BH services to indigenous populations as directed in stability and reconstruction operations, humanitarian assistance, disaster relief, and peace support operations. The comprehensive support provided by the CSC medical company entails all of the nine COSC functional mission areas. The COSC functional mission areas are identified in Chapter 1 and discussed in subsequent chapters of this publication.

CAPABILITIES

2-24. At TOE Level 1, the CSC medical company provides—

- Advice, planning, and coordination for COSC to commanders.
- Reconstitution (COSC) support for units up to division size.
- Preventive and fitness teams (four to ten personnel) for consultation, treatment services, and reconstitution support for up to battalion-sized organizations.
- Soldier restoration or reconditioning programs for up to 50 Soldiers per fitness team on an area basis.
- Deployment of COSC elements to forward areas for support of contingency operations.

ASSIGNMENT

2-25. The medical company, CSC (TOE 08467A000) is assigned to a Medical Command (US Army) (MEDCOM). Elements of this TOE may be further attached to a corps MEDBDE or to either a MMB or an ASMB.

ORGANIZATION

2-26. The medical company, CSC is organized into a headquarters section, a preventive section, and a fitness section. The company is dependent on appropriate elements of the MEDCOM or MEDBDE for administrative and MEDLOG support, medical regulating, COSR casualty delivery, and medical evacuation. The company is dependent on appropriate elements of the corps or EAC for finance, legal, personnel and administrative services, food service, supply and field services, supplemental transportation, and local security support services. When CSC medical company elements or teams are deployed to division areas, they are dependent on the division medical companies (such as MSMC or FSMC/BSMC) for patient accounting, transportation, food service, and field service support.

Employment

2-27. The medical company, CSC is employed in all intensities of conflict when a corps with two or more divisions is deployed. Task-organized COSC elements are deployed for division-sized combat operations, stability and reconstruction operations, and other contingency operations which are METT-TC-dependent. Together, the preventive and fitness sections provide all five BH disciplines. These resources are flexibly task-organized in a variety of combinations to meet the fluid COSC threat at different phases in the operations. Personnel may be quickly cross-attached from one section to another to accommodate the shifting workload and to provide reconstitution support packages. The preventive and fitness sections both organize into teams. The COSC preventive or fitness teams deployed forward of the corps boundaries in support of tactical operations come under the control of the FHP operations element in the supported units. These teams will also come under the operational control of the division or brigade CSC teams. One or more of the medical company, CSC's eight preventive teams may locate at the BSMC when deployed in DS of SBCT, HBCT, and separate brigades or ACRs.

Fitness Teams

2-28. One or more of the four (ten-person) fitness teams may reinforce ASMCs which are deployed to locations throughout the corps and EAC. These teams provide a basis for COSC prevention and intervention. The teams may conduct Soldier restoration programs at the ASMCs, as required. These teams may also be deployed forward to provide temporary augmentation/reinforcement, as required.

2-29. Based on workload, one or more of the four fitness teams, plus one or more preventive teams locate with a theater hospital where they conduct Level IV COSC reconditioning programs, as required. A hospital located in the corps rear or out of the CZ is the best location to conduct the theater COSC reconditioning program. When deployed with a hospital, these teams provide mobile consultation in the vicinity of the hospital. These teams are also prepared to restrict reconditioning programs and deploy forward in support of higher priority missions on very short notice. These teams can also augment hospital BH ward services by staffing a temporary BH ward.

2-30. The medical company, CSC is divisible into four functionally emulative increments for split-based operations and stability and reconstruction operations as assigned. Nonstandard task elements for specific missions can be organized using any combination of the preventive section and fitness section personnel to meet specific BH needs. For stability and reconstruction operations, the minimum is an officer/NCO team to supplement a brigade COSC team or a preventive module/team of two officers, one NCO, and one enlisted. These modules may be augmented with personnel from the fitness section to add additional specialty expertise.

Headquarters Section

2-31. The headquarters section provides C2 and unit-level administrative and maintenance support to its subordinate sections when they are collocated with the company. The headquarters section may also provide assistance to detached elements by making site visits if the elements are within a feasible distance for ground transportation. The CSC medical company elements normally deploy with limited maintenance capability. When these COSC elements deploy, they are dependent on the supported units for patient accounting, transportation, food service, and field services. The personnel assigned to the headquarters section are identified in Table 2-5. See Appendix C for information on the AOE L-Series TOE medical company and medical detachment, CSC.

Table 2-5. Medical company, combat stress control, headquarters section personnel

Personnel
Commander (LTC, AOC 60W00, MC)
Chaplain (CPT, AOC 56A00, Chaplain [CH])
Field Medical Assistant (CPT, AOC 70B67, MS)
First Sergeant (E8, MOS 68W50)
Mental Health NCO (E7, MOS 68X40)
Chemical Operations Specialist (E5, MOS 54B20)
Patient Administrative NCO (E5, MOS 68G20)
Supply Sergeant (E5, MOS 92Y20)
Power Generator Equipment Repairer (E4, MOS 52D10)
Wheeled Vehicle Mechanic (E4, MOS 63B10)
Administrative Specialist (E4, MOS 71L10)
Patient Administrative Specialist (Two)—(E4 and E3, MOS 68G10)
Armorer (E4, MOS 92Y10)
Cook (E3, MOS 92G10)
Note. Personnel from the headquarters section are deployed with teams or task-organized COSC element, as required.

SECTION II — OPERATIONAL STRESS ASSESSMENT TEAM IN A THEATER

MISSION OF OPERATIONAL STRESS ASSESSMENT TEAMS

2-32. The operational stress assessment team (OSAT) is task-organized when needed from personnel in the research institutes of United States Army Medical Research and Materiel Command (USAMRMC). The mission of this team is to deploy with BH research expertise, equipment and supplies to provide scientifically valid needs assessment of multiple units in the theater and macro analysis of COSC stressors, stress, and Soldier needs. The source is predominantly the Walter Reed Army Institute of Research, augmented when needed with research psychologists and enlisted technicians from the US Army Aeromedical Research Laboratory, the US Army Medical Research Institute of Chemical Defense, and the US Army Research Institute of Environmental Medicine. The OSATs are USAMRMC teams that are assembled entirely from personnel in table of distribution and allowances (TDA) units and are dependent on TOE units when they deploy to a theater.

ASSIGNMENT AND CAPABILITIES

2-33. The OSAT normally is used for augmenting the senior medical command in the theater. The OSAT has the capabilities to—

- Plan, coordinate, and conduct questionnaire and interview surveys.
- Establish a database and analyze scientific-quality data from their own surveys and from standardized unit assessments conducted by unit MH section and CSC units for their local supported units.
- Provide rapid reports of findings to the command that include recommendations for policy and for COSC actions and interventions.
- Design and validate survey instruments, including ones to address new and emerging stressors and issues in the theater.
- Use USAMRMC capability to archive and analyze data for comparison with other conflicts.
- Educate and mentor COSC and other personnel in survey methodology and provide them, as well as the theater commander, the wider perspective to use in their local UNA.

- Interface with the theater surveillance capabilities of preventive medicine (PVNTMED) and occupational and environmental medicine to address the BH aspects of chemical, biological, radiological and nuclear (CBRN) exposures and threats.

DEPENDENCY

2-34. The OSAT depends on the medical headquarters or the units with which it is working for administrative support. The OSAT depends on the theater (line) command for implementation of the approved surveys.

MOBILITY AND SECURITY

2-35. The OSAT depends on the medical and theater command and the units with which they work for transportation. The OSAT personnel are issued personal weapons and protective equipment, but depend on the units they are with for protection.

COMMUNICATIONS

2-36. The OSAT depends on the medical and theater command and the units with which they work for communications. With approval of theater command, they may be allowed USAMRMC-issued phones for satellite communication for mission-essential purposes.

ORGANIZATION

2-37. The typical OSAT has three officers and two enlisted Soldiers. The officer positions are listed as research psychologists (AOC 71H), but may be filled by clinically qualified BH officers with research training and experience. The enlisted are usually MH specialists (MOS 68X) with research training and experience. Staffing may also include personnel with information management specialties. The equipment is the required personal equipment of the team members, plus data collection and automated data processing equipment appropriate to the mission.

Chapter 3

Combat and Operational Stress Control Support Operations

SECTION I — COMBAT AND OPERATIONAL STRESS CONTROL PROFESSIONAL DISCIPLINES AND PROFESSIONAL CONSULTANTS

PROFESSIONAL DISCIPLINES

3-1. There are five BH professional disciplines and two enlisted specialties that serve in support of the COSC mission. The professional disciplines and enlisted specialties were identified in Chapter 2. The professional disciplines include social work, clinical psychology, psychiatry, occupational therapy (OT), and psychiatric/BH nursing; the enlisted specialties are in BH and OT. While much of the COSC knowledge base and most of the skills are shared by all the BH personnel, each discipline brings its own perspective from its professional training, skills that can only partially be familiarized to the others, and in some cases unique credentials to conduct specific assessments and treatments. The special perspectives and skill sets of the officer and enlisted members of the COSC team are summarized in the following subparagraphs.

OFFICER AREAS OF CONCENTRATION

Social Work Officer

3-2. The social work officer (AOC 73A67) evaluates the psychosocial systems affecting the Soldier, his family, and the unit; identifies and coordinates with military and civilian support agencies (including the Army Substance Abuse Program and the Family Advocacy Program [FAP]); assesses BH disorders; provides a range of counseling and psychotherapies. Installation MTFs are responsible for BH services that are beyond the scope of COSC functional areas such as long-term clinical care, inpatient services, and alcohol treatment programs. This officer provides proactive consultation, conducts individual and group counseling, supervises Soldier restoration/reconditioning, and coordinates RTD of recovered cases. He also provides staff advice and coordinates Army and civilian social services support.

3-3. When deployed as a member of a COSC preventive team or task-organized COSC element, the social work officer's duties include—

- Evaluating psychosocial (unit and family) functioning of Soldiers with COSR and misconduct stress behavior.
- Coordinating and ensuring the return of recovered COSR and neuropsychiatric (NP) Soldiers to duty and their reintegration into their original or new unit.
- Identifying and resolving organizational and social environmental factors which interfere with combat readiness.
- Coordinating support for Soldiers and their families through Army and civilian community support agencies, when possible.
- Apprising unit leaders, primary care physicians, and others health care providers of available social service resources.
- Providing consultation to supported unit commanders and to other BH/COSC personnel regarding problem cases.
- Counseling and providing therapy for Soldiers with psychological problems.
- Conducting and supervising unit needs assessment and TEM.

3-4. The doctoral level social work officer, as outlined in DODD 6490.1, may conduct clinical evaluations for imminent risk or dangerousness.

Clinical Psychologist

3-5. A clinical psychologist (AOC 73B67) diagnoses BH disorders and a range of psychotherapies including cognitive/mental approaches, assessment for safety evaluations, and psychometric assessment.

Psychiatrist

3-6. The psychiatrist (AOC 60W00) is a physician that is credentialed to diagnose and treat BDPs. He also conducts mental assessments, makes differential diagnosis, provides a range of psychotherapies and safety evaluations, and prescribes medication.

Psychiatric/Behavorial Health Nurse

3-7. This AN officer (AOC 66C007T) provides specialized nursing services for emotionally distressed individuals and promotes COSC within the MTF and the adjacent military community. He performs liaison and consultative functions to ensure continuity of patient care. The additional skill identifier of 7T indicates that advanced skills and competencies are required for this position and that the individual may act as a consultant in clinical nursing practices.

Occupational Therapist

3-8. The occupational therapist (AOC 65A00), plans, implements, and supervises preventive and restorative OT services to enhance the occupational performance of Soldiers. This officer evaluates Soldier's performance across the spectrum of occupational areas (such as activities of daily living, work, education, leisure, and social participation). Analyze jobs and job tasks for underlying performance requisites. Identify and evaluate mental and physical stressors and stress reactions and teach prevention, adaptive coping and psychosocial skills. Apply therapeutic media in both, individual and group settings to include environmental adaptation, the therapeutic use of self, consultation, education, and occupation (purposeful activity). Match Soldiers to therapeutic and/or vocational activities based on therapeutic need and functional ability.

ENLISTED MEDICAL OCCUPATIONAL SPECIALTIES

Mental Health Specialist

3-9. The MH specialist (MOS 68X30/20) provides the perspective of the enlisted Soldier, takes the initial history and conducts mental status evaluations for supervising officer, and administers psychological tests. The MH specialist also performs counseling and may have advanced specialty training (such as drug and alcohol counseling with higher ranked individuals).

Occupational Therapy Specialist

3-10. The OT specialist (MOS 68W30/20N3) provides the perspective of the enlisted Soldier, assists the supervising OT officer in evaluating a Soldier's occupational performance; conducts initial occupational performance history interviews and mental status evaluations; observes clients to gather data as part of task performance skill assessments; and implements OT interventions under the supervision of an OT officer.

CONSULTANTS

3-11. There are two levels of COSC consultants—the theater-level consultant and the subordinate commander consultant. The theater-level COSC consultant is normally the senior COSC officer designated by the theater surgeon and approved by the theater commander. The subordinate command COSC consultants are MH officers that function as the point of contact (POC) for one or more commanders.

These consultants oversee, coordinate, and conduct COSC interventions and activities for their respective commanders. They accomplish the following tasks appropriate to their level of operation:

- Direct and report AO-wide COSC needs assessments.
- Coordinate AO-wide implementation of COSC functional areas.
- Ensure COSC standards are established and consistently applied.
- Ensure compatibility between COSC standards and hospital BH standards (theater/AO COSC consultant).
- Define COSC operational requirements.
- Recommend the AO COSC evacuation policy (theater/AO COSC consultant).
- Prepare the COSC portion of the FHP estimate and operations plan.
- Review all command medical policies affecting COSC interventions and activities.
- Evaluate the quality of COSC interventions and activities rendered in the AO.
- Make regular AO COSC consultant site visits to COSC sections and units.
- Coordinate with other COSC consultants in adjoining AOs and in the primary medical support elements to ensure unity of effort.
- Oversee AO COSC training.
- Coordinate joint and combined COSC interventions and activities as necessary.
- Plan for future COSC operations.
- Consult with the command surgeon on all COSC matters.

SECTION II — DEPLOYMENT AND EMPLOYMENT OF MENTAL HEALTH SECTIONS

MENTAL HEALTH SECTIONS

3-12. The MH section assignments, staff, functions and responsibilities were discussed in Chapter 2. The information provided in this section pertains to deployment and employment and how these sections prepare for COSC support for their units in garrison, combat operations, and stability and reconstruction operations predeployment. The involvement of MH section with support units during field and other training exercises is important. It permits BH personnel to gain a familiarity with type of units supported, how they conduct operations, their mission, and the likely stressors associated with a particular type of unit. It also allows them to interact and become familiar with unit personnel and to gain some level of trust through their interactions. Combat and operational stress control personnel should be active participants in all training exercises. Unit predeployment training must be planned when the unit is alerted for its deployment but scheduled after cross-leveling and filling of all positions is completed. Training of BH personnel will include—

- Ensuring their active participation in the normal training cycle of the BCT/ACR to which assigned.
- Training the BH officers and specialists in all COSC functional areas that support their unit.
- Working with the supported units on a frequent basis so BH personnel may gain a familiarity with the unit's personnel, capabilities, and possible stressors associated with its mission.
- Providing predeployment train-up to include refresher and operation-specific training before deployment.

PLANNING

3-13. Since planning is normally performed at battalion level, MH sections are involved in developing a COSC estimate to support the unit plan. The COSC estimate of the situation is derived from mission analysis, to include considerations for COSR, BH disorders, substance abuse, misconduct behaviors, suicide, major stressors, troop populations to be supported, and COSC augmentation requirements or assets available.

COORDINATING

3-14. The BH personnel may be involved in coordinating COSC support with supporting and supported units, as appropriate. Brigade BH personnel must be known and trusted by the leaders of the units they support. Brigade BH personnel coordinate with other BH officer and enlisted personnel pertaining to the professional disciplines associated with conducting the COSC mission. They also coordinate with other key medical and nonmedical personnel (such as health care providers, surgeons, chaplains, UMTs that assist or may be involved with the COSC mission).

DEPLOYMENT OF MENTAL HEALTH SECTIONS

Early Deployment

3-15. The MH section should be deployed early into staging areas with the lead elements of the medical company and supported units.

Command and Control

3-16. The MH section falls under the C2 of the FSMC/BSMC and may receive mission guidance from the brigade surgeon through the FSMC/BSMC commander. Tasking for COSC operational support flows through the BSMC/FSMC headquarters. When BH personnel from the medical detachment/company, CSC are attached, the MH section ensures the timely flow of operations and situation awareness information to that element.

EMPLOYMENT OF THE MENTAL HEALTH SECTIONS

3-17. Employment of the MH section is based on the type of operations being conducted. The MH section normally operates from the FSMC/BSMC located in the brigade support area (BSA). When MH personnel are deployed forward, they are dependent on supported units for rations and logistical support.

MENTAL HEALTH FUNCTIONAL AREA CONSIDERATIONS

3-18. When determining COSC support for BCT units, the BH team should review each of the BH functional areas and determine which BH functions need to be implemented for each supported unit. Also, in developing the COSC plan, the BH team should consider—

- Conducting a UNA for each supported battalion and separate company supported.
- Establishing a consultation and education/training program for supported units to include—
  - Traumatic event and transition management.
  - The need for COSC triage, stabilization, and BH treatment should be addressed as they arise.
  - Soldier restoration, which can only be accomplished within and by the BSMC holding section or with augmentation by or referral to CSC unit personnel.
  - Reconditioning services that will need to be referred to nearest location where available.

SECTION III — SUPPORT OPERATIONS CONDUCTED BY THE MEDICAL UNIT, COMBAT STRESS CONTROL

PREDEPLOYMENT ACTIVITIES FOR ALL MEDICAL UNITS, COMBAT STRESS CONTROL

3-19. The predeployment activities identified in Section II for MH sections also apply for the units in Section III that include the ASMB or MMB and medical companies and detachments, CSC. These units must also keep themselves focused on applying the nine functional areas during the predeployment phase.

Note. The MMB is the C2 organization which will replace the ASMB, the medical logistics battalion, and the evacuation battalion.

MEDICAL DETACHMENT, COMBAT STRESS CONTROL DEPLOYMENT

3-20. The medical detachment, CSC is deployed to provide COSC support whenever a division or two BCTs/regiments are deployed. It is assigned to a medical brigade or other medical headquarters for C2. Elements of the detachment may be further attached to FSMCs/BSMCs for operational control and logistical support while in support of the BCT.

MEDICAL COMPANY, COMBAT STRESS CONTROL DEPLOYMENT

3-21. The primary mission of this unit is to provide COSC support throughout a corps AO. It is assigned to either a MEDCOM/MEDBDE and may support from two or up to five divisions. They assign teams to any operational area based on troop populations without BH support. They move teams rapidly to areas of surge workload or mass casualties.

MEDICAL FORCE 2000 VERSES THE MEDICAL REENGINEERING INITIATIVE DETACHMENTS

3-22. The Medical Force 2000 (MF2K) and MRI CSC detachments conduct COSC support operations very similarly. However, in the MRI detachment, psychiatry assets are assigned to the fitness section and psychology assets are assigned forward to the prevention section as compared to the MF2K detachment, the manning strategy for psychology and psychiatry is reversed. The psychologists are assigned to the fitness section and psychiatrists are assigned to the prevention teams.

Medical Detachment, Combat Stress Control Support to a Division and Brigade Combat Teams During Combat Operations

3-23. When in support of an AOE division, a medical detachment, CSC is usually attached to the MSMC/BSMC and for a transformation division the medical detachment, CSC is attached to the MMB or MEDBDE. It can be attached to other battalion-sized units for support. It is under the operational control of the support battalion and MSMC/FSMC/BSMC and closely coordinates its augmentation support with the division surgeon and division psychiatrist. Habitual long-term relationships of medical detachments, CSC with specific divisions should be established so that COSC support is integrated as a normal function of the FHP mission. However, as an echelon above division (EAD) asset, the detachment (or its modular teams) may be task-organized and deployed to support other units or missions other than divisional units, as workloads require. The medical detachment, CSC depends on the units to which it is attached for administrative and logistics support. It may be reinforced and/or receive personnel replacements from a medical company, CSC or from another medical detachment, CSC.

Preventive Section Operations

3-24. Upon arrival to a theater, the detachment is initially attached to a higher medical headquarters for C2 and its three CSCP teams of four-person each are usually further attached. One CSCP is attached to each divisional MSMC prior to commencement of combat operations. This will allow the teams to linkup with and augment division COSC assets in support of the brigades.

- The deployment of each CSCP provides additional COSC personnel that increase COSC augmentation capabilities at the FSMC. A CSCP may routinely visit the BASs to provide consultation. The CSCP may deploy, as needed, to provide reconstitution support to units undergoing hasty or deliberate reorganization.
- In some operations, some or all of the CSCP personnel may either be attached to an ASMC or are deployed in support of a maneuver brigade or reconstitution site. It is important for all CSCP personnel to understand that they are responsible for all functional areas, and will conduct them as dictated by the mission.

Fitness Section Operations

3-25. The CSCF may locate with any Level II or above MTF in the AO. This section has tents and is equipped to operate a Soldier restoration center and can supervise Soldier restoration activities to include consultation, rest, relaxation, and treatment as necessary for speedy RTD within three days. The CSCF also provides triage and stabilization at the MSMC and consultation to other division units located in the DSA and division rear areas. All CSCF members are responsible for providing support for all COSC functional areas as dictated by the mission.

3-26. The CSCF staff may move within the AO to temporarily reinforce or reconstitute a CSCP at the brigade level, or to escort COSR casualties to the Soldier restoration center. The CSCF personnel give reconstitution support to attrited units, especially when the units return to the DSA. The CSCF supplements the division psychiatrist; brigade COSC teams, CSCPs, and chaplains and leaders in after-action debriefings. The CSCF helps integrate recovered Soldiers and new replacements into units during reorganization activities. When either the BSA or DSA is tactically too unstable to allow Soldier restoration operations, the CSCF may locate with an ASMC or a CSH and attempt to continue their support to the division from that location.

COMBAT STRESS CONTROL DETACHMENT IN CONTINGENCY OPERATIONS

3-27. For contingency operations, a task-organized CSC element from the CSC medical detachment may be deployed. A CSCP supporting a maneuver brigade may operate out of a central base of operations. The team may deploy forward to the BSA (base camps or fire bases), in response to—

- Anticipated battle.
- After-action debriefing requirements.
- Alcohol/drug problems in a unit.
- Incidents of misconduct stress behaviors.
- High incidence of COSRs.
- Unit rotation in or out of theater.

3-28. Several CSCF and CSCP from two or more CSC medical detachments may be consolidated under the command of the senior medical detachment, CSC commander to staff a central reconditioning program for the AO. This may also function as an alcohol/drug detoxification rehabilitation program. This may be referred to as a COSC center and will also provide consultation and treatment support to MP confinement facilities where misconduct stress behaviors may have led to incarceration. This could also include support for EPW and detainee confinement facilities.

STABILITY AND RECONSTRUCTION OPERATIONS

3-29. In stability and reconstruction operations, if the force deployed is smaller than a division, a medical detachment, CSC would not be required to provide COSC support. In such cases, the medical company, CSC or the medical detachment, CSC may be tasked with providing either a COSC team or a task-organized COSC element. The COSC team or the task-organized COSC element is attached to the supporting medical headquarters or to an MTF and conducts its mobile consultation mission.

MEDICAL UNITS, COMBAT STRESS CONTROL INTERFACE AND COORDINATION REQUIREMENTS

3-30. The medical units CSC must interface with its higher headquarters element and with the unit to which it is attached. The higher headquarters may be a MEDCOM, MEDBDE, or an ASMB/MMB.

- Interface between the CSC units and their higher headquarters. The medical unit CSC interfaces with its higher headquarters pertaining to its assigned mission. It provides estimates and has input to the operation plan (OPLAN). The unit receives its OPORD from the higher headquarters. Interface between the unit and higher headquarters staff elements will include the following subject areas—

- The COSC operations.
- Assignment or attachment of the medical detachment, CSC elements.
- Daily personnel and equipment status reports.
- Class VIII (medical supply) status and resupply requirements.
- Casualty feeder reports.
- Operation orders.
- Personnel replacement for the detachment.
- Medical intelligence information.
- Behavioral health/COSC consultation tasking and results.
- Maintenance requirements and requests.
- Replacement and reconstitution operations.
- Civil-military operations.
- Communications (signal operation instructions [SOI], access to message centers and nets, and transmission of COSC messages through medical, land, and other channels).
- Mass casualty plan.
- Road movement clearances.
- Tactical updates.
- Contingency operations.
- Return-to-duty and non-RTD procedures.
- Medical evacuation procedures (air and ground ambulances).
- Changes in locations of supported unit.

- Interface and coordination with unit to which attached. The headquarters of the unit to which a medical detachment, CSC is under operational control (OPCON) or attached is responsible for providing the administrative and logistical support requirements of the detachment. These requirements are normally identified in the attachment order. The higher headquarters will be identified in the attachment order and the medical detachment, CSC will be coordinated with this headquarters prior to deployment. The medical detachment, CSC must coordinate with the headquarters staff according to the tactical standing operating procedures (TSOP) of the unit of attachment. The staff shares information with the detachment commander or his representative pertaining to the threat, tactical situation, patient/COSR casualty status, and changes in FHP requirements. Coordination activities and subject area information exchange should include—
  - Command and control procedures.
  - Status of FSMCs/BSMCs and CSCPs.
  - Communications and SOI.
  - Operational support requirements.
  - Civil-military operations.
  - Soldier restoration operations.
  - Reinforcement and personnel replacement.
  - Road movement and clearances.
  - Casualty reporting and accountability.
  - Patient-holding procedures.
  - Force protection.
  - Convoy operations.
  - Geneva Conventions.
  - Detainees/EPW.
  - Improvised explosive devices (IEDs).
  - Religious and cultural considerations.

RECONDITIONING CENTERS

3-31. The medical company, CSC task-organizes CSC elements to staff separate small reconditioning centers in locations that are relatively secure. However, under some circumstances, the company may consolidate teams to establish a large reconditioning center, which supports two or three divisions. Reconditioning facilities normally locate near a CSH. See Chapter 12 for definitive information on reconditioning center operations.

Chapter 4
Unit Needs Assessment

SECTION I — PRINCIPLES AND TENETS OF UNIT NEEDS ASSESSMENT

GENERAL PRINCIPLES

4-1. The UNA is a systematic process for identifying the COSC needs of units. The UNA allows COSC personnel to identify priorities for interventions and activities and for allocating resources. The UNA is not a clinical screening to identify individuals who have or are at risk for BH disorders problems, but rather evaluates the needs of the Soldier population and leads to more effective preventive COSC activities and early interventions. The UNA allows COSC personnel to—

- Identify and describe specific areas of COSC need.
- Discover factors contributing to the needs.
- Provide an assessment of the BH training needs of Soldiers, leaders, UMT, and medical personnel within the unit.
- Develop plans to meet or improve the COSC needs of Soldiers and units through prevention and early intervention activities.

SCOPE

4-2. Unit needs assessments can be conducted at various command levels from small to large units. The COSC personnel at each level should conduct UNAs for their supported units. Larger-unit UNAs can include the composite findings and recommendations of one or more UNAs completed at subordinate levels. For example, a brigade UNA may include the results of several subordinate battalion UNAs (consolidated for confidentiality). Generally, UNAs are not conducted below the company level, though exceptional circumstances may dictate a platoon or lower UNA.

TENETS

4-3. Tenets of UNAs—

- Gain commander approval and support prior to conducting the UNA.
- Protect anonymity and ensure confidentiality of Soldiers and commanders. This includes the protection of unit identification from higher headquarters.
- Provide the commander with an unbiased assessment.
- Consider the social, political, and organizational factors of the environment.
- Ensure that information sources represent the entire unit.
- Select an assessment method that is consistent with the operational situation.
- Limit overgeneralizing the findings from one unit to another or from one time or situation to another.
- Recognize that UNAs provide population-level assessments of COSC needs, not clinical screening tools to identify individuals who may benefit from COSC interventions.
- Distinguish between what respondents report they need and what interventions are required.
- Ensure the UNA is planned and coordinated at a level commensurate with the complexity of the assessment and/or situation before starting.

SECTION II — FOCUS AND METHODS OF DETERMINING UNIT NEEDS ASSESSMENT

AREAS OF FOCUS FOR A UNIT NEEDS ASSESSMENT

4-4. A UNA involves the systematic assessment of numerous areas of Soldier and unit functioning. A typical UNA includes, but is not limited to, the following areas:

- Major stressors impacting the unit.
- Level of unit cohesion.
- Well-being of unit Soldiers.
- Soldier concern about home-front issues.
- Soldier knowledge and skill for controlling combat and operational stress.
- Soldier ideas for addressing COSC needs.
- Soldier knowledge of accessing COSC resources.
- Barriers and stigma that prevent Soldiers from accessing COSC services.
- Training needs of Soldiers, leaders, UMT, and medical personnel on topics of COSC importance (such as buddy aid, suicide awareness, or suicide prevention). See FM 6-22.5, FM 22-51, and US Army Training and Doctrine Command (TRADOC) Pamphlet 600-22 (available at: http://www.tradoc.army.mil/tpubs/pamndx.htm) for additional information on suicide awareness and prevention.

METHODS OF UNIT NEEDS ASSESSMENTS

4-5. The UNA takes advantage of all available information. Various methods can be used to assess general unit needs and to identify issues that differ among subgroups (for example, gender, rank, or race/ethnicity). The use of multiple assessment methods is recommended. These methods may include—

- Interviewing Soldiers to hear their perceptions and concerns. Information gathered during the casual conversation, although informal, may afford valuable anecdotal information (as well as developing trust and familiarity).
- Reviewing policy documents (standing operating procedures [SOPs]).
- Interviewing key unit personnel (chain of command, chaplain, and medical personnel).
- Conducting structured group interviews (focus groups or unit survey interviews).
- Administering standardized surveys and questionnaires (paper/pencil or web-based).
- Monitoring trend indicators (such as high rates of BH referrals, sick call, or misconduct, Soldier suicide, sexual assault, fratricide or disciplinary actions).
- Using multiple methods for gathering information when possible to ensure different viewpoints are considered.

Planning Considerations for Unit Needs Assessment

4-6. The UNA varies in complexity and formality depending on the purpose of the assessment and the needs of the supported unit commander. Many factors determine the complexity and formality of a UNA, which in turn influence its feasibility.

Complexity

4-7. Complexity equates to the requirements and cost to complete a UNA. Complexity is influenced by factors such as—

- The size and number of units to be assessed.
- Geographic dispersion of the units and time constraints.

Formality

4-8. Formality equates to the degree that scientific principles and methods are employed in the conduct of the UNA. An example of a UNA with low formality is one where the assessing team uses nonstructured group and individual interviews and perhaps brief questionnaires they have developed. The information is obtained from key individuals and a convenience sample of troops. That may be sufficient to quickly identify problems and make recommendations to resolve them. A UNA with high formality is one that uses professionally validated questionnaires and structured focus group interviews according to standard protocols. The data is collected from a scientifically selected sample of the larger population. The data is analyzed by standard analysis programs, so that statistically selected samples of valid comparisons can be made with other similarly sampled units in other geographical locations and across time. The OSAT expertise and assistance may be necessary for some UNA instruments and methods. In general, increasing formality increases the complexity of the UNA. The UNA requires a higher level of formality as the need for objective data and scientific precision increases (such as when results from multiple units are to be merged or compared with other UNAs).

Feasibility

4-9. Feasibility is the ability to accomplish a UNA with available resources. In developing a particular UNA, trade-offs are made to achieve an acceptable level of data quality (formality) for an affordable cost (complexity).

UNIT NEEDS ASSESSMENT PROCESS

4-10. All UNAs are conducted following a three phase plan.

PREASSESSMENT

4-11. Preassessment is an initial phase to obtain command support, determine target issues, and select appropriate methods to use.

ASSESSMENT

4-12. Assessment is the phase for gathering, integrating, and analyzing information to identify the COSC needs of the unit.

POSTASSESSMENT

4-13. The principal task of the postassessment phase is to determine the COAs to present to the commander which address the identified COSC needs. These findings are then linked to a plan of action.

WHEN TO CONDUCT UNIT NEEDS ASSESSMENTS

4-14. Unit needs assessments may be conducted at various times throughout the deployment cycle. A UNA may be conducted—

- Prior to initiation of COSC interventions and activities while COSC personnel and units establish their support relationships.
- To assess the effectiveness of COSC interventions and activities that are in the process of being conducted or that have been completed.
- At the request of supported commanders.
- To monitor trend indicators (see Paragraph 4-5).
- After serious traumatic events and significant unit transitions.
- To collect unit information for COSC planning.

4-15. Unit needs assessments should be conducted during all types of deployments, including stability and reconstruction operations, and combat operations.

Chapter 5
Consultation and Education

SECTION I — PRINCIPLES AND TENETS OF CONSULTATION AND EDUCATION

GENERAL PRINCIPLES

5-1. Combat and operational stress control consultation is defined as the transmission of information through an interactive relationship between the consultant and consultees. Education is used here in a broader sense in that it is the transmission of information by any means. Examples of consultation include providing COSC advice, coaching, training, and planning assistance. During this process, the consultant learns about the consultees and their needs, and tailors the interactions accordingly. Examples of education include distributing flyers, video and radio broadcasts, and news articles. The consultation and education functional area supports the other COSC functional areas.

CONSULTANTS

5-2. The term BH consultant is used in the general sense to describe any person performing COSC consultation or education. In addition to those theater or subordinate commander COSC consultants identified in Chapter 3, all BH/COSC personnel may serve as consultants at their level. Familiarization training with other BH disciplines enriches the ability to serve as a consultant.

CONSULTEES

5-3. Depending on the units in the area of support, broad types or categories of personnel may be consultees. These personnel are involved in recognition and control of stress as a result of their position or duty assignments. Consultation and education consultees may include—

- The command or unit surgeon and his staff.
- Staff chaplain and UMT.
- The senior commander and the senior NCO of a battalion, brigade, division, or corps.
- Staff officers and NCOs, including adjutant and personnel (S1/G1), intelligence (S2/G2), operations (S3/G3), civil-military affairs (G9), and the Judge Advocate General (JAG).
- Medical personnel such as PVNTMED teams that have missions that often complement the COSC mission.
- Company grade leaders, especially company commanders, executive officers, first sergeants (1SGs) platoon leaders, platoon sergeants, company NCOs, and Soldiers that are trained to be peer mentors.

SOLDIER-PEER MENTORS

5-4. Soldiers selected by their commanders may be trained to provide COSC help-in-place assistance for COSC information to peers. They may also serve as a POC between fellow Soldiers and the COSC and UMT teams. Selected Soldier-peer mentors with additional training could assist commanders with conducting COSC training in mission risk assessments.

TENETS FOR CONSULTATION AND EDUCATION

5-5. Consultation and education are ongoing processes that are performed across the deployment cycle and the continuum of operations. They may be provided in response to a specific request by command or recommendation of COSC personnel. Consultation and education may be provided during routine scheduled meetings such as a commander's weekly update or a leader professional development class.

5-6. Effective consultation is accomplished by the consultant's active outreach. Consultation is best conducted through recurring face-to-face contact, preferably at the consultee's location. Telephone and radio may be used to setup initial meetings and provide follow-up consultation. Consultation may be conducted one-on-one or in small groups where interaction is feasible. When necessary and feasible, audio or video teleconference may suffice. Active outreach supports the functional areas particularly COSC triage and traumatic event and transition management.

5-7. Successful consultation depends on the consultant's credibility, and the trust and familiarity established with consultees. In addition to sound professional knowledge base and clinical skills, the consultant must have military bearing and knowledge of the military (including the units, missions, vocabulary, acronyms, and skills involved). Rapport is enhanced by the demonstration of the consultant's genuine interest in the consultee and the unit. There are some situations when consultation and education are more effectively provided by the COSC officers, the NCOs, or the junior enlisted personnel.

SECTION II — CONSULTATION, EDUCATION, AND PLANNING

CONSULTATION PROCESS

5-8. The following six steps outline the process for most consultations:

- Initiate the process by introducing yourself and your capabilities to key leadership and get approval to continue.
- Assess the needs of consultees and formulate ways to address them.
- Present COA to consultee and define goals and feasibility of alternative actions.
- Implement approved COA.
- Evaluate outcomes or progress.
- Plan follow-up actions.

5-9. Consultation and education encompasses a broad range of topics, extending from prevention to treatment. The process of consultation and education assists the consultee with anticipation, identification, and control of stressors (environmental, physiological, cognitive, and emotional) and stress reactions (adaptive and maladaptive stress reactions). The following list of COSC functions, resources and their availability provides examples of consultation and education topics:

- Operation risk factors.
- Individual risk factors.
- Stress moderators.
- Unit cohesion, esprit de corps, and morale.
- Critical event management.
- Barriers to care and overcoming stigma.
- Operations planning.
- Integration of new personnel to unit.
- Trust in equipment and supporting units.
- Tough realistic training.
- Home-front stressors.
- Indicators of unit stress level.
- Combat and operational stress behaviors.
- Combat and operational stress reactions.

- Misconduct stress behaviors.
- Mental disorders.
- Suicidal and homicidal behaviors.
- Command referral processes.
- Scapegoating behaviors.
- Rumor control.
- Physical needs (such as sleep, nutrition, and hygiene).
- Social/emotional needs (such as morale, welfare, recreation push package).
- Communication with home.
- Reintegration of Soldiers into unit.
- Alcohol and substance abuse.
- Procedures for COSC triage.
- Care provider stress (such as compassion fatigue, vicarious trauma, moral dilemma).
- Long-term consequences of COSR and mental disorders.
- Medical evacuation policy and procedures.
- Leadership.
- Fear management.
- Rest and relaxation (R&R) considerations.

5-10. Additional details about some of these topics may be found in FM 22-51.

5-11. From theater staff level down to the MH sections, BH personnel are involved in the consultation and planning process for COSC support. Their support may be either through providing situational updates, developing staff estimates, or being directly involved and assisting with the development of the FHP plan. Planning starts with mission analysis. The mission analysis is the first step of the military decision making in abbreviated planning for a time-constrained situation, see FM 5-0. A part of mission analysis is based on the commander's intent and guidance. A medical battalion headquarters staff may receive addition instructions from the MEDBDE headquarters staff. During the planning process, command surgeons and medical planners (S3/G3) may seek information from the BH/COSC consultant or task medical company/detachment, CSC for COSC estimates.

MISSION ANALYSIS

5-12. The COSC planning is an essential part of every medical estimate and operations plan. Ensure that the command surgeon involves the COSC consultant in all medical planning. Mission analysis includes—

- Assessing COSC capabilities (organic and attached assets with current status and location).
- Gathering information from the commander's guidance regarding the upcoming operations. (See the S2 for intelligence, the S3 for scheme of maneuver and forces involved; the S1 for casualty estimates, personnel status issues, and units supported, and the logistics staff officer [S4/G4] for logistical considerations). Surgeons provide disease and nonbattle injuries (DNBI) estimates and status of medical units. Unit needs assessments, organic medical personnel, and UMT can provide additional information.
- Developing the COSC estimate based on the gathered information. The COSC estimate helps medical planners to anticipate demands on medical and COSC resources and to prepare accordingly. See FM 3-0, FM 5-0, and FM 8-55 for methods to develop the COSC estimate.
- Assessing limitations (specify reasons that BH assets are not available).
- Identifying specified, implied and essential COSC task in the FHP annex of the OPORD.
- Providing COSC COA.
- Presenting estimate and COA to command surgeon. Define goals and feasibility.
- Implementing approved COA. Consult with implementing COSC personnel about estimate and operational details.

- Evaluating outcomes or progress.
- Providing estimate and COA for fragmentary order (FRAGOs) and planning follow-up actions.

REHEARSAL

5-13. To achieve optimal synchronization, the FHP plan is rehearsed as an integral part of the combined arms plan at the combined arms rehearsal. Medical leaders provide input to the over all plan and develop the concept for the FHP plan. During the decision-making/orders process, they identify critical events and synchronize the FHP plans. In addition to medical locations on the CSS overlay, these plans indicate the triggers for FHP events. At the rehearsals, leaders practice their synchronized plans that include FHP.

5-14. The CSS/FHP annex of the OPORD that includes map overlays is the conclusion of the medical planning efforts; the rehearsal is the culmination of the preparation phase for an operation. The medical platoon leader has the responsibility for rehearsing FHP operations. Rehearsals are done to achieve a common understanding and a picture of how the plan will be implemented. The following leads to a successful rehearsal—

- All plans must be completed prior to the rehearsal.
- The FHP portion of the battalion rehearsals should focus on the events that are critical to mission accomplishment. A successful rehearsal ensures explicit understanding by subordinate medical personnel of their individual missions; how their missions relate to each other; and how each mission relates to the commander's plan. It is important for all medical units to understand the complete FHP concept.
- Rehearsing key FHP actions allows participants to become familiar with the operation and to visualize the "triggers" which identify the circumstances and timing for friendly actions. This visual impression helps them understand both their environment and their relationship to other units during the operation. The repetition of critical medical tasks during the rehearsal helps leaders remember the sequence of key actions within the operation and when they are executed.
- The OPORD is issued through effective troop leading procedures.

SECTION III — TRANSITION MANAGEMENT AND SUPPORT IN THE DEPLOYMENT CYCLE

SUPPORTING TRANSITION AND PHASES OF A DEPLOYMENT

5-15. Transition and phases of a deployment cycle may have different and sometimes unique stressors. These stressors may be subjects for consultation and education. Needs assessment may be necessary to redistribute COSC assets and change the means for delivering COSC activities and interventions.

5-16. The COSC personnel conduct transition (change of command) workshops, especially after relief of commanders. These workshops are normally requested by the incoming commander. The purpose of these workshops is to—

- Facilitate staff discussion of what the staff sees as the unit and staff's strong points and the areas needing more work.
- Provide the new commander the opportunity to discuss his leadership style and his expectations and set priorities for the staff.

5-17. The change of OPTEMP may include—

- Going from deployment to combat (offensive or defensive operations).
- Conducting battle handover in place by turning the battle over to another unit.
- Going from high intensity combat to stability and reconstruction operations.

PREDEPLOYMENT SUPPORT

5-18. Medical support in predeployment is mainly the responsibility of MEDDAC personnel or of a US Army Reserve (USAR) mobilization site augmentation unit (which has no BH personnel assigned).

Division and brigade COSC BH personnel are involved with their unit's Soldiers who are being treated for mental disorders, and with general COSC prevention of the unit members and families. The CSC units that are not deploying may be tasked to support predeployment COSC activities and interventions. Combat and operational stress control screening of individuals during predeployment may include—

- Secondary BH screening for Soldiers who are referred by primary care providers that flagged positive on DD Form 2795 (Pre-Deployment Health Assessment).
- Health screening of medical and BH records.
- Fitness for duty evaluations according to AR 40-501.
- Evaluations and recommendations to leadership about Soldiers who are fit for duty overall, but should not be deployed at this phase of the operation for BH reasons.

5-19. Other predeployment functions may be provided on request. These functions may include—

- Briefings, consultation and education on deployment cycle stressors and how to cope, and on specific stressors that Soldiers may encounter in the AO.
- Unit assessments at request of commanders.

MID-TOUR REST AND RECUPERATION OR EMERGENCY LEAVE

5-20. Leaders must be aware and alert for Soldiers who exhibit a need for COSC screening prior to mid-tour R&R. Other Soldiers may only require some education briefings and handout cards on handling their transition from a hostile CZ to R&R environments. In some cases, prior to emergency leave, some screening for domestic violence risk may also be necessary in any potential high-risk Soldier. This COSC screening is done for ensuring the safety of the Soldier and of others. One protocol has unit commanders referring those Soldiers that have requested leave for the purpose of addressing marital, legal, or other highly charged difficulties at home, to the BH officer. The BH officer explores the situation and the Soldier's reactions to it and checks for history with the rear detachment of the MH section, MEDDAC/MEDCEN and FAP. If the BH officer finds the Soldier to be at high risk, the commander can deny leave. Another option is to escort the Soldier to home base. At home base, he is kept by his unit's rear detachment until a meeting is held under safe conditions with the Soldier, BH, FAP personnel, his family, and the others involved in the difficulty. At this meeting the situation is defused or resolved before the Soldier is permitted to leave the rear detachment area.

REDEPLOYMENT

5-21. For Soldiers, redeployment is the process of getting orders to redeploy, embarking towards the demobilization site, or deploying to another out-of-the-theater mission. Screening of these individual Soldiers may be a requirement and have established criteria that should be used to ensure appropriate screening is conducted. The theater or specific units may require that all primary care providers who perform screening interview, postdeployment health assessment have a designated BH consultant. Available COSC/BH personnel provide secondary one-on-one BH screening for Soldiers who are referred by primary care providers because they flagged positive on DD Form 2795. Some Soldiers may require brief treatment or referral may be required as a result of the screening.

END-OF-TOUR STRESS MANAGEMENT

5-22. End-of-tour stress management is essential to reduce mental problems with Soldiers returning to their home station and/or families. This also promotes Soldiers seeking help early when problems occur. Mental health/COSC personnel consult with command on activities during redeployment and for postredeployment. Issues may include—

- Common mental disorders and potential misconduct may result when Soldiers are relieved of the stress and the focus of dangerous missions and are looking towards and preparing to go home.
- Memorial services for those Soldiers who died during the deployment.
- Recognition of outstanding performance and the equitable awarding of decorations.
- Closure ceremonies.
- Homecoming-reunion education briefing and training.

5-23. The end-of-tour debriefings (EOTDs) are like leader-led after-action debriefings, except that the time period covered extends from predeployment through to this point in redeployment. The EOTD should be conducted while the unit is still in the theater or at some place where transportation home pauses for one or more days. For each phase of the operation, the participants review what were the significant events and the problems or stressor that bothered the group (including critical events, if any). Equally important the group recalls what went well, how problems and stressors were overcome (at least by some members), and what positive memories and feelings they had and will take home.

FACILITATORS

5-24. The COSC or UMT facilitators should be invited to advise the leader and sit in if there are likely to be many negative memories and feelings that have disrupted or threaten unit cohesion. Advise leaders and troops on the importance of sustaining unit identity and contact with teammates beyond postdeployment and in the future.

POSTDEPLOYMENT

5-25. Postdeployment covers that period of time from embarking from theater, through scheduled postdeployment activities, and return to work after block leave. As with predeployment, medical support in postdeployment is mainly the responsibility of installation MTF, often augmented by a USAR mobilization site augmentation unit. Division and brigade COSC/BH personnel and CSC units that are themselves demobilizing as their units are deactivating will have limited involvement with the BH and COSC needs of fellow Soldiers. These personnel continue to provide, as needed, consultation to the command and to key personnel of the unit that they were assigned/attached. They consult with the installation MTF about BH and coordinate COSC needs that should be met. The CSC units that were not deployed may be tasked to support postdeployment COSC activities and interventions. The COSC personnel who are not demobilizing provide reintegration education. Screening of individuals postdeployment will continue and secondary screening will continue in postdeployment and be recorded on a DD Form 2796 (Post-Deployment Health Assessment). Each primary screener should have a designated BH consultant from the MEDCEN/MEDDAC. Secondary screening and treatment of referrals are provided by the installation MTF. Recent research with a returning division indicated that a secondary postdeployment screening from 90 and 120 days after returning from the theater may be necessary. This seems to be the time when persisting or delayed symptoms, problems, and perhaps impairment show need for intervention. All Soldiers should be advised about follow-up BH support resources. Soldiers that are leaving active service should also be informed about deployment-related entitlements and benefits.

READINESS PHASE OF THE DEPLOYMENT CYCLE

5-26. Division/brigade MH sections and medical units, CSC, of Active Army and Reserve Component (RC) must use their training opportunities to the fullest. During these training exercises, they should provide COSC consultation, education, and training to supported units the same as they would in an actual deployment. Mental health personnel will perform many of their critical mission functions and should not just simulate them. The management of real COSR occurs in field and garrison/home-station settings. Triage and emergency stabilization may be required. Restoration may be appropriate in some field exercises. Even reconditioning can sometimes be provided as a multiday "course" at Active Army posts where CSC units are stationed. Further, TEM and transition management are also potential major roles for organic MH sections and medical units, CSC in peacetime. The RC as well as Active Army medical units, CSC have been used extensively to respond to traumatic events and to assist units that are in the premobilization, mobilization, or postdeployment phases of their deployment cycles.

Chapter 6
Traumatic Event Management

OVERVIEW OF TRAUMATIC EVENT MANAGEMENT

TAILORING TRAUMATIC EVENT MANAGEMENT

6-1. Traumatic event management blends other COSC functional areas to create a flexible set of interventions specifically focused on stress management for units and Soldiers following potentially traumatizing event (PTE). Like other functional areas, COSC providers must tailor COSC and TEM support to the needs of the unit and Soldier.

6-2. An event is considered *potentially traumatic* when it causes individuals or groups to experience intense feelings of terror, horror, helplessness, and/or hopelessness. Guilt, anger, sadness, and dislocation of *world view* or faith are potential emotional/cognitive responses to PTEs. Studies of Soldiers in Operation Iraqi Freedom (OIF) and Operations Enduring Freedom (OEF) have shown a correlation between exposure to combat experiences and BH disorders, most particularly acute stress disorder and PTSD. The following events should be monitored as PTEs for Soldiers and units:

- Heavy or continuous combat operations.
- Death of unit members due to enemy or friendly fire.
- Accidents.
- Serious injury.
- Suicide/homicide.
- Environmental devastation/human suffering.
- Significant home-front issues.
- Operations resulting in the death of civilians or combatants.

Scope of Traumatic Events Management

6-3. In this chapter, TEM is discussed as it applies to military units and personnel to support readiness. Traumatic events management can be adapted to nonmilitary groups or individuals brought together by a PTE. Traumatic events debriefing should be conceptualized as an ongoing process and not an acute intervention. For military units and personnel, TEM is active in all phases of the deployment cycle and across the continuum of military operations.

Traumatic Events Management Functional Mission Area

6-4. Successful TEM relies on a solid foundation of other COSC functional area activities such as UNA that may require the COSC provider to—

- Establish a credible working relationship with supported unit leadership.
- Understand the unit's needs via previously conducted UNAs.

6-5. In the absence of preexisting relationships and UNA, COSC providers should secure command support and recommend a UNA as the first step in TEM. Subsequent UNAs will clarify unit responses to other TEM interventions and to ongoing unit needs.

Unit Needs Assessments

6-6. When conducting a UNA after a PTE the COSC providers need to take under consideration the following—

- Ensure a timely arrival that does not disrupt unit operations but facilitates the UNA.

- They should not interrupt or intrude on the people who are attending to the acute crisis when arriving at the unit, unless asked. Presence without intruding will gain trust information and POC.
- Understanding that the UNA may be limited by the urgency of the unit's return to action, the difficulty of data collection, and having limited resources. Sufficient knowledge must be gained to tailor the interventions to the unit before initiating them. It is better to defer the intervention to the next opportunity if the unit must return to action immediately after replenishment and/or before necessary data is collected.
- Know that the UNA is a unit-level assessment and does not substitute for individual-level screenings or COSC triage. See Chapter 4 for a detailed discussion of UNA.

Consultation and Education

6-7. The COSC provider should conduct unit leader consultation and education activities prior to a PTE. Consultation and education topics should include—

- The impact of PTE on unit and Soldier readiness.
- Common PTEs for units and Soldiers.
- An overview of TEM.
- Components of TEM.
- Normal responses to PTEs.
- Triggers to refer Soldiers for BH evaluation.
- Development of SOPs for responding to PTEs when they occur.

6-8. Preemptive consultation and education prepare unit leaders to institute TEM interventions following a PTE. The 5 R's are a good model to build on. In acute TEM interventions, leaders should consider interventions that target—

- Safety, security, and survival.
- Food, hydration, clothing, and shelter.
- Sleep.
- Medication (replace medications destroyed or lost).
- Orientation of unit/Soldiers to developing situation.
- Restoration of communication with unit, dependents, friends, and community.
- Location should be at a secure site that provides protection from ongoing threats, toxins, and harm.

6-9. Consultation and education to Soldiers should emphasize normalizing the common reactions to trauma, improving their coping skills, enhancing self-care, facilitating recognition of significant problems, and increasing knowledge of and access to COSC services. Post-PTE UNAs guide further consultation and education efforts. Combat and operational stress control providers should be aware that leaders may not have experienced the PTE in person, but have experienced the PTE through their Soldiers and may require support.

6-10. For an in-depth discussion of consultation and education, refer to Chapter 5.

Combat and Operational Stress Control Triage

6-11. The COSC provider should be prepared to provide COSC triage in the aftermath of the PTE. Consultation and education is important to ensure appropriate and timely referrals for triage. The COSC provider can offer the following guidelines in referral of Soldiers for COSC triage:

- Persistent or worsening traumatic stress reactions (such as dissociation, panic, autonomic arousal, and cognitive impairment).
- Significant functional impairments (such as role/work relationships).
- Dangerousness (suicidal or violent ideation, plan, and/or intent).

- Severe psychiatric comorbidity (such as psychotic spectrum disorder, substance use disorder, or abuse).
- Maladaptive coping strategies (such as pattern of impulsivity or social withdrawal under stress).
- New or evolving psychosocial stressors.
- Poor social support.
- Failure to respond to acute supportive interventions.
- Exacerbation of preexisting psychiatric conditions.
- Soldier request for assessment.

Stabilization

6-12. The COSC provider should be prepared to provide or coordinate stabilization services following the PTE. Pre-PTE coordination with medical unit personnel promotes safe management.

Soldier Restoration

6-13. The measures below are applicable to Soldiers with COSR following a PTE. The COSC provider should be familiar with the 5 R's and with BICEPS. In keeping with restorative efforts, the COSC provider focuses on the following measures through leadership consultation, Soldier education, and/or direct management:

- Minimizing exposure of Soldiers with COSR to further PTE.
- Reducing physiological arousal.
- Mobilizing support for those who are most distressed.
- Providing information and fostering communication and education.
- Using effective risk communication techniques.
- Proving assurance/reassurance.
- Mitigating fear and anxiety.
- Encouraging sleep hygiene.
- Reestablishing routines.
- Promoting exercise and nutrition.
- Encouraging self-paced emotional ventilation.
- Discouraging use of alcohol/substances.

Behavioral Health Treatment

6-14. Given the correlation between PTE and development of traumatic stress disorders the COSC provider must be familiar with the best current practices for evaluation and treatment. The Veterans Health Administration (VHA)/DOD Clinical Practice Guidelines website (http://www.oqp.med.va.gov/cpg/cpg.htm) offers clinicians evidence-based assessment and treatment algorithms for acute stress disorder, PTSD, and many other BH disorders.

6-15. In recent years, the use of early interventions in response to PTE has come under critical review. The focus of much of this debate is on the use of psychological debriefing (PD) and more specifically, the critical incident stress debriefings provided to individuals or groups exposed to PTEs. Current research suggests that PD/critical incident stress debriefings can be harmful to participants, while failing to reduce traumatic stress reactions or to prevent the progression to PTSD. As PD/critical incident stress debriefings undergo definitive study, COSC providers should adhere to current evidence-based best practices.

6-16. In accordance with the VHA/DOD Clinical Practice Guidelines (http://www.oqp.med.va.gov/cpg/cpg.htm) for PTSD, COSC providers should—

- Consider alternative methods to PDs for individuals affected by PTEs.
- Avoid PD as a means to reduce acute posttraumatic distress (acute stress reaction or acute stress disorder) or to slow progression to PTSD.

- Understand there is insufficient evidence to recommend for or against conducting structured group debriefings.
- Be aware that compulsory repetition of traumatic experiences in a group may be counter-productive.
- Consider group debriefings with preexisting groups (such as teams, units, emergency medical treatment [EMT] teams, coworkers, family members) may assist with group cohesion, morale, and other important variables that have not been demonstrated empirically.
- Emphasize that group participation must be voluntary.

LEADER-LED AFTER-ACTION DEBRIEFING

6-17. A leader-led after-action debriefing is lead by a platoon, squad, or team leader and is not normally conducted above platoon level. The leader-led after-action debriefing should be conducted after all missions especially when the maneuvers did not go according to plan. A leader-led after-action debriefing may even be sufficient for PTEs involving injury or death. For the leader to conduct a PD, his personnel should have received previous PDs; and normally provide peer support and validation for showing and talking about their emotional reactions during and after the debriefing. The best time to conduct this debriefing is as soon as is feasible after the team/squad/platoon has returned to a relatively safe place and members have replenished bodily needs and are no longer in a high state of arousal. Usually a well-conducted leader-led after-action debriefing is the best option to manage PTEs during a mission. The exception to this type of debriefing is when the event evoked reactions that seriously threaten unit cohesion and/or have a high likelihood of arousing disruptive behavior and emotions. In these situations the leader should ask himself the following—

- Should I conduct the debriefing?
- Should a trained facilitator be present?
- Should a request for COSC TEM be submitted for his team/squad/platoon?

CONDUCTING A LEADER-LED AFTER-ACTION DEBRIEFING

6-18. These debriefings require the leader to extend the lessons-learned orientation of the standard AAR. He uses the event reconstruction approach or has the individuals present their own roles and perceptions of the event, whichever best fits the situation and time available. See Training Circular (TC) 25-20 for definitive information on AARs. When individuals express or show emotions, the leader and the teammates recognize and normalize them; they agree to talk about them later and support the distressed Soldier through personal interactions. The group then returns to determining the facts. Lessons-learned discussion is deferred until all the facts are laid out. See FM 22-51 for additional information. The leader may provide education about controlling likely reactions or referral information at the end, depending on his knowledge and experience.

6-19. When a PTE is likely to create individual or collective guilt, distrust, or anger, the unit leader should be encouraged to request COSC assistance. Either a COSC or a UMT person trained in TEM sits in with the leader-led debriefing as a familiar and trusted friend of the unit. The COSC or UMT facilitator helps the unit/team leader rehearse and mentors the leaders on the debriefing process. During the PD, the facilitator can ask questions of the group to clarify the facts and steer the process away from divisive anger, blaming, and scapegoating toward a positive, cohesion-restoring outcome. This method is halfway between a simple leader-led after-action debriefing and a PD and is referred to as a *facilitated leader-led after-action debriefing*. The leader conducting the debriefing must be attentive to identify individuals needing COSC follow-up.

6-20. Leaders in positions above platoon level also have a role in leader-led after-action debriefing. Company commanders and 1SGs may conduct leader-led after-action debriefing with their subordinate leaders. Battalion commanders and higher may conduct leader-led after-action debriefings with their staff after distressing actions and may include subordinate leaders when time allows bringing them together.

COOL DOWN MEETINGS

6-21. A cool down meeting is referred to as an immediate, short meeting when a team or larger unit/group returns from the battlefield or other missions. These cool down meetings are held after heavy/intense battles with the enemy or a shift in the mission has occurred which is highly arousing and/or distressing. This is especially important after critical events. The cool down meeting is an informal event and occurs before the participants fully replenish their bodily needs and precedes any other activities including more COSC interventions, or return to the mission.

6-22. Personnel who coordinate and wait for the return of the unit that has been in a heavy intense battle, include leaders or supporting officers or NCOs from the command, UMT, and COSC providers. (In domestic support operations the same personnel identified above or other trained personnel from governmental or nongovernmental organizations such as the Red Cross may be waiting on the return of the unit after a PTE. These personnel may be present at the cool down meeting.)

COMPONENTS OF A COOL DOWN MEETING

6-23. Components of a cool down meeting may include—

- Assembling all of the unit personnel at a safe and relatively comfortable location for a brief period of time (about 15 minutes).
- Receiving or sharing nonstimulating beverages and convenient food (*comfort* foods if available).
- Providing personnel the opportunity to talk among themselves.
- Giving recognition and praise for the difficult mission they have completed.
- Providing information to unit personnel on where and how they will rest and replenish.
- Previewing the immediate agenda for the unit on what will happen after the cool down meeting including plans for further debriefing and/or other available stress control or morale/welfare intervention.
- Providing announcements pertaining to further preparations and expected time of return to the mission.

6-24. The COSC personnel may have a role in cool down meetings. In consultation and education, COSC consultants emphasize the value of cool down meetings and the simplicity of the components, which are easily neglected in crisis situations. When feasible, they unobtrusively attend the cool down meeting, showing "presence" while learning of the event, getting familiar with the key people, and observing anyone showing signs of distress and being available to them. If requested, they may give a very brief introduction and review of normal stress reactions that unit personnel may have in the next few hours.

Chapter 7

Combat and Operational Stress Control Support for Units Undergoing Reconstitution

UNIT RECONSTITUTION SUPPORT

7-1. Unit reconstitution support is defined in FM 100-9 as extraordinary actions that commanders plan and implement to restore units to a desired level of combat effectiveness commensurate with mission requirements and available resources. Besides normal support actions, reconstitution may include—

- Removing the unit from combat.
- Assisting the unit with external assets.
- Reestablishing the chain of command.
- Training the unit for future operations.
- Reestablishing unit cohesion.

7-2. When tasked as part of a reconstitution task force, COSC personnel are responsible for providing units with Soldier restoration, performing the COSC functions of the UNA, and providing consultation and education, as required. They provide triage, stabilization, Soldier restorations, and short time-constrained COSC treatment, when needed. Reconstitution is a time-constrained process, but Soldier restoration may be provided at the reconstitution site if several days are available. The focus of consultation and education efforts include—

- Rebuilding unit cohesion.
- Integrating new Soldiers or groups of Soldiers into the unit.
- Facilitating assumptions of command by replaced leaders.
- Facilitating the building of Soldiers' confidence in their leaders and themselves.
- Mentoring unit leaders in leader-led after-action debriefing process.
- Advising on COSC aspects of bereavement memorial services and communication with the family support group and unit families. The COSC personnel also facilitate leader-led after-action debriefings and lead or serve as observers in TEM interventions.

RECONSTITUTION PROCESS

7-3. Reconstitution of units transcends normal day-to-day force sustainment actions. Reconstitution is a total process. Its major elements are reorganization, assessment, and regeneration, in that order. All COSC personnel should be thoroughly familiar with FM 100-9.

Reorganization

7-4. Unit reorganization primarily involves a shifting of internal resources and is accomplished as either immediate or deliberate reorganization. The commander of the attrited unit decides to reorganize when required and may consider the following—

- Immediate reorganization is the quick and usually temporary restoring of degraded units to minimum levels of effectiveness. Normally, the commander implements immediate reorganization at his combat location or as close to that site as possible to meet near-term needs. The COSC personnel provide consultation with key POC in the unit (leaders, UMTs, and trauma/health care specialist, usually by telecommunication).
- Deliberate reorganization is done to restore a unit to the specified degree of combat effectiveness. Usually, more time and resources are available further to the rear. Procedures are similar to immediate reorganization except that some personnel and weapons system

replacement resources may be available, equipment repair is more intensive, and more extensive cross- leveling is possible. When used in reorganization, cross-leveling involves the movement of personnel and/or equipment between units to achieve equalization. The process is accomplished while maintaining or restoring the combat effectiveness of the units involved.

- The role of COSC personnel in deliberate reorganization may require a COSC team to deploy to the reorganization site. They assess the stressors and stress reactions and advise the commanders on supportive actions, (such as those for the reorganization of small unit-level elements), assists command with transitions and integration of new replacements.
- When the reorganization involves battalion size or larger units the process becomes more like the coordinated operation described for unit regeneration, provided below. It is likely to draw upon fitness team expertise and Soldier restoration assets from rearward as well as on the forward deployed teams.

REGENERATION

ASSESSMENT PHASES

7-5. The defining characteristic of regeneration is the massive infusion of personnel, equipment, and assistance under the directions of higher headquarters. Assessment measures a unit's capability to perform its mission. It occurs in two phases. The unit commander conducts the first phase. He continually assesses his unit before, during, and after operations. If he determines it is no longer mission-capable even after reorganization, he notifies his commander. The higher headquarters either changes the mission of the unit to match its degraded capability or removes it from combat. External elements may also have to assess the unit after it disengages. This is the second phase. These elements do a more thorough evaluation to determine regeneration needs. They also consider the resources available.

7-6. The second phase of assessment begins with an initial survey by a team sent by the higher headquarters. This team determines the status and needs of the attrited and exhausted unit as it moves to the regeneration site. Some of the key issues in estimating the COSC needs of the unit include the—

- Percentage and nature of casualties.
- Duration of operations and environmental exposure.
- Status of hydration, nutrition, and sleep.
- Loss and current effectiveness of leaders.
- Attitudes, perceptions, and level of confidence of unit survivors.
- Traumatic events.

7-7. The COSC consultants at every level must provide consultation to all command surgeons on the importance of including BH/COSC personnel in all reconstitution operations. The initial evaluation team should include a subordinate commander COSC consultant. Task-organized CSC teams normally deploy to the unit during the second phase to provide UNA, consultation, and other COSC activities and interventions. Unit needs assessment with high formality may be requested by the higher headquarters for decision-making purposes. These assessments are feasible for selected units, with prior coordination, as much of the logistical complexity is reduced by the orchestrated planning at higher command echelons.

Conducting Regeneration

7-8. Regeneration involves rebuilding a unit through large-scale replacement of personnel, equipment, and supplies. If required, it includes internally reorganizing; reestablishing or replacing essential C2 and conducting mission-essential training for the reconstituted unit. Regeneration is required when heavy losses of personnel and equipment leave a unit combat-ineffective and unable to continue its mission. The commander controlling assets to conduct regeneration decides whether to use resources for this purpose. Regeneration has two variations: incremental regeneration or whole-unit regeneration. Incremental regeneration is the massive infusion of individual personnel replacements and single items of equipment into the surviving unit elements. Whole-unit regeneration is the replacement of whole units or definable subelements, such as squads, crews, and teams. The S1/G1, S4/G4, and medical staffs coordinate the dispatch of the regeneration task-force teams. These teams occupy the reconstitution site before arrival of the exhausted unit. The COSC assets that perform the indicated initial assessment are needed as part of this task force. The reconstitution task force guides each element of the arriving units into its designated areas. The regeneration task force provides for the immediate needs of the survivors. This should include personal gear and equipment to replace lost or damaged items, food services, personnel services, maintenance teams, and medical teams to provide sick call services while organic medical personnel rest. Replacement personnel are sent to the reconstitution site. The COSC personnel assist with their assimilation into the regenerated unit.

RECONSTITUTION RESOURCE REQUIREMENTS FOR COMBAT AND OPERATIONAL STRESS CONTROL

7-9. Factors which influence the resource requirements for COSC in reconstitution (deliberate reorganization and regeneration) include the size of the unit, number of subunits which have suffered heavy casualties, the extent of emotional trauma, and time available. The focus of COSC for reconstitution support was provided in Paragraph 7-2. A guideline is provided in Table 7-1 for COSC personnel requirements. This staffing should not overly rely on organic COSC staff that will be in need of rest.

Table 7-1. Reconstitution operations guideline for combat and operational stress control personnel requirements

Size of Unit	*Personnel Required*
Company/Troop	2 to 4
Battalion/Squadron	6 to 12
BCT/Brigade Regiment	12 to 30
Division	30 to 60

Chapter 8
Combat and Operational Stress Control Triage

SECTION I — THE TRIAGE PROCESS FOR COMBAT AND OPERATIONAL STRESS CONTROL

SOLDIER TRIAGE

8-1. The COSC triage process is the sorting of Soldiers based on an assessment of their needs and capabilities, and the location where they can best be managed in keeping with BICEPS principles. Triage is applicable at every level of care. The two key components of COSC triage are assessment and disposition.

8-2. Assessment is an evaluation of the Soldier's physical and BH needs, potential medical emergencies, and other safety risks. Assessment is performed by COSC personnel according to their professional training, expertise, and standards.

8-3. Disposition is the COSC intervention plan to address the needs identified in the assessment. Disposition has two components that include—

- Determining what intervention techniques best address the Soldier's needs and functional capabilities.
- Selecting the best location where the Soldier can be managed. The personnel conducting COSC triage should consider the needs, abilities, and the safety of the Soldier. They should also consider the unit's capacity to provide COSC interventions based on its OPTEMPO mission, resources, response to prior consultations, and willingness to participate in COSC interventions.

TRIAGE ALGORITHM FOR COMBAT AND OPERATIONAL STRESS CONTROL

8-4. Like surgical triage categories, COSR also uses triage categories. Each of the COSR triage categories are discussed in detail in the below paragraphs. The COSC triage algorithm presented in Table 8-1 uses some of the triage categories.

Table 8-1. The combat and operational stress control triage algorithm

<table>
<tr><td>Step 1</td><td colspan="5">Is this a medical emergency?</td></tr>
<tr><td></td><td>Yes</td><td>Refer to nearest MTF</td><td>No</td><td colspan="2">Go to Step 2</td></tr>
<tr><td>Step 2</td><td colspan="5">Does the Soldier require medical/behavioral observation?</td></tr>
<tr><td rowspan="3"></td><td rowspan="3">Yes</td><td rowspan="3">Go to Step 3 A</td><td rowspan="3">No</td><td colspan="2">Does Soldier have presumptive COSR or MH disorder?</td></tr>
<tr><td>Yes</td><td>Go to Step 3 B</td></tr>
<tr><td>No</td><td>Help-in-place</td></tr>
<tr><td>Step 3</td><td colspan="2">A. Can the Level II MTF or CSC Soldier Restoration Center provide adequate evaluation and intervention?</td><td colspan="3">B. Can the Soldier's unit support the 5 R's or other treatment interventions?</td></tr>
<tr><td rowspan="4"></td><td>Yes</td><td>HOLD*</td><td>Yes</td><td colspan="2">UNIT</td></tr>
<tr><td rowspan="3">No</td><td rowspan="3">REFER</td><td rowspan="3">No</td><td colspan="2">Is there a suitable support unit?</td></tr>
<tr><td>Yes</td><td>REST</td></tr>
<tr><td>No</td><td>HOLD*</td></tr>
<tr><td colspan="6">* When deciding between two or more potential Level II MTFs or CSC Restoration Centers, refer the Soldier to the one closest to his unit that meets his COSC needs.</td></tr>
<tr><td>Step 4</td><td colspan="5">Use on subsequent triages. Has the Soldier improved after appropriate duration of intervention?</td></tr>
</table>

TRIAGE CATEGORIES FOR COMBAT AND OPERATIONAL STRESS REACTION CASES

8-5. The following are triage categories that may be used for COSR cases. *Help-in-place* (HIP), *rest*, *hold*, and *refer* cases are discussed below.

HELP-IN-PLACE CASES

8-6. *Help-in-place* is used to identify those cases that do not have severe COSR or BH disorders. They are provided COSC consultation and education, as appropriate, and remain on duty. These interactions may occur in any setting (for example, dining facility, workplace, or the post exchange). Individual identifying information is not retained or documented. There is no implicit or explicit therapist-patient or therapist-client relationship in HIP interactions.

8-7. The unit identifies those cases that remain with or return to their original unit, either for full duty with their section/platoon or for light duty with extra rest and replenishment within a headquarters element. This option depends on the unit's mission, resources, and the Soldiers symptoms. Personnel performing triage must, therefore, be familiar with the unit's situation and take that into account. When the Soldier's condition improves, the Soldier and/or unit may not feel that additional triage is necessary.

Note. *Help-in-place* is a new triage category that replaces the old *duty* category and is also used for unit cases.

REST CASES

8-8. *Rest* identifies those cases that are provided rest and replenishment in a nonmedical support unit, usually one that is in support of their unit. These Soldiers do not require close medical or BH observation or treatment. They are unable to return immediately to their own unit either because their unit cannot provide an adequate environment for the 5 R's; or transportation is not available for at least a day; or the 5 R's can best be coordinated from the nonmedical support unit. This option depends on the resources and mission of the available CS/CSS units, as well as on the Soldier's symptoms. Someone must be designated to be in charge of the Soldier and ensure that the 5 R's are provided. There must be a reliable transportation link to return the Soldier to his original unit after a day or two. When the Soldier's condition improves sufficiently for him to return to his unit, the Soldier and/or the supporting unit may feel that additional triage is necessary.

HOLD CASES

8-9. *Hold* refers to those cases that require close medical/BH observation and evaluation at a Level II MTF or COSC Soldier restoration center because the Soldiers symptoms are potentially too disruptive or burdensome for any available CSS unit or element. Soldier's symptoms may be caused by a BH disorder that could suddenly turn worse and require emergency treatment. The Level II MTF or COSC Soldier restoration center must have the capability to provide the necessary medical observation, diagnostic tools, and adequate stabilization for emergency treatment. When deciding among capable Level II MTFs or COSC Soldier restoration centers, refer the Soldier to the one closest to his unit that meets his COSC needs. Assessment of closeness considers speed and reliability of transportation and back. Consider transferring to another Level II MTF or COSC Soldier restoration center with increased capabilities before changing a Soldier's triage category to *refer*. All *hold* cases will be triaged again by COSC personnel or other trained medical personnel after they have been placed in this category.

REFER CASES

8-10. *Refer* cases are similar to the *hold* cases, except that *refer* cases are too disruptive and burdensome for the MTF or the COSC Soldier restoration center that is not resourced to care for this particular case. The MTF or COSC Soldier restoration center cannot provide the necessary level of diagnostic and treatment capabilities. *Refer* cases requiring care at a COSC reconditioning center, a Level III MTF or higher levels of care will be triaged by COSC or other trained medical personnel prior to being transferred to these facilities.

TRIAGE PERSONNEL FOR COMBAT AND OPERATIONAL STRESS CONTROL

8-11. All COSC personnel participate in the triage process according to their professional training, experience, and standards. Familiarization training among BH disciplines extends the effectiveness of all COSC personnel in triage skills. Medical care providers must be mentored to use the COSC triage process. A commander providing the 5 R's and placing a Soldier in a support unit for a temporary break does not equate to COSC triage.

TRIAGE CONSIDERATIONS

8-12. Triage should be initiated when the—

- Soldier is a self-referral.
- Chaplain has referred the Soldier.
- Medical personnel have requested a COSC consultation and referred the Soldier.

- Unit member/buddy has referred the Soldier.
- Leader has requested an informal referral.
- Soldier is a command-directed referral (see DODD 6490.1 and DOD Instruction [DODI] 6490.4).
- Combat operational stress control personnel observe a Soldier's behaviors which indicate possible COSR or a BH disorder.

8-13. Factors that influence an assessment may vary in depth and duration due to several other factors. These factors may include the—

- Referral source.
- Nature of the complaint.
- Observed needs.
- Medical/psychiatric history.
- Availability of resources.
- Amount of privacy for conducting assessment.
- Operational Tempo.
- Environmental conditions.
- Professional training of the person making the assessment.
- Command interest.
- Soldier cooperation.

Note. Regardless of these factors, COSC personnel are responsible for conducting the assessment in a timely manner within professional standards of practice.

DOCUMENTATION

8-14. An assessment is documented according to AR 40-66 whenever the Soldier—

- Is diagnosed with a BH disorder.
- Has a condition (or suspected condition) requiring emergency medical evaluation or treatment.
- Is prescribed medication.
- Is assessed to be a high safety risk (for example, homicidal or suicide, cognitive impairment, substance abuse, and impulsivity).
- Requests that documentation of his assessment be made in his medical records.
- Is evacuated beyond Level II MTF for further assessment or treatment.
- Is command-referred for a BH evaluation.

TRANSFER AND EVACUATION

8-15. All COSC personnel are responsible for knowing the transfer/evacuation policies and procedures within their AO. Policies and procedural information are available through the command surgeon, medical regulating officer (MRO) of the medical command, control, communications, computers, and intelligence (C4I) headquarters or major MTF. All relevant background and/or clinical documentation must accompany the Soldier during the transportation or evacuation process.

MODE OF TRANSPORTATION

8-16. Nonambulance transport is the preferred mode of transportation for COSR and nonurgent BH cases. Examples of nonambulance transportation include the Soldier's unit vehicles, supporting supply/logistics vehicles, and nonambulance medical vehicles. Ambulances convey patient status on Soldiers and often must be reserved for medical emergency. Under the provisions of the Geneva Conventions, ambulances must be used exclusively in the performance of humanitarian duties, therefore, they cannot be used to return Soldiers to duty and transport must be provided by the supported unit.

ESCORT

8-17. Depending on a Soldier's condition, an escort (either medical or nonmedical) may be necessary to provide safety, monitoring, and accountability during transportation or evacuation. The escort should be an NCO or officer of equal or greater rank/grade as that of the escorted Soldier. Escorts must be emotionally mature, responsible, and capable of conducting their escort duties. Frequently escorts carry the Soldier's clinical documentation to the destination MTF.

FEEDBACK

8-18. Good communication is essential for effective continuity of care during the transportation/evacuation process. The Soldier's unit must be informed about his location and status throughout the process. The sending party must provide sufficient documentation about the Soldier's condition, history, and administered interventions. The receiving party must provide feedback to the sending party regarding receipt of the Soldier and his documentation.

SECTION II — PRECAUTIONS AND DIFFERENTIAL DIAGNOSTIC PROBLEMS ASSOCIATED WITH COMBAT AND OPERATIONAL STRESS CONTROL TRIAGE

PRECAUTIONS FOR COMBAT AND OPERATIONAL STRESS CONTROL TRIAGE

8-19. Medical emergencies must be identified during COSC triage. Medical emergencies consist of physical illnesses or injuries and/or BH disorders that can result in permanent injury, disability, or death. Early identification of a medical emergency avoids unnecessary delay in definitive treatment. Medical emergencies can cause emotional and/or BH changes and may resemble COSR in presentation. The following conditions and behaviors could be medical emergencies. Therefore, it is important to ensure medical examinations and disposition of Soldiers that display the following conditions—

- Psychosis.
- Mania.
- Alcohol withdrawal.
- Substance intoxication.
- Delirium.
- Suicidal gesture, attempt, or high risk for suicidal behavior.
- Catatonia.
- Significant paresis, paralysis, and/or sensory loss.

8-20. Assessing for physical illnesses or injuries is a critical part of COSC triage. The COSC personnel must always consider physical illnesses or injuries that resemble COSR or BH disorders. Physical illnesses or injuries may not reach the threshold of a medical emergency, but must be recognized and appropriately treated. Assessing for physical illnesses or injuries requires an adequate review of body systems and a quick physical examination. The examination includes vital signs, examination of head, eyes, ears, nose, throat, chest, abdomen, and extremities with simple testing of reflexes and muscle strength. In field situations (Level I and II), negative or normal findings need to be documented on DD Form 1380 (US Field Medical Card). Refer to AR 40-66 for detailed information on patient accountability and management of individual health records. Any positive findings from the physical examination must be evaluated further. If the examiner has not checked various body systems, it is not reassuring to tell a Soldier that his physical or BH complaints are only COSR. All COSC personnel should receive familiarization training on basic medical examination techniques and in documenting medical information. Whenever a physical illness or injury is suspected, personnel should consult with their medical peers for further assistance. Some cases will require direct medical examination by a physician or physician assistant (PA). The COSC personnel should not order tests or procedures that do not directly influence case management. Medical tests may promote the patient role in the mind of the Soldier. Needless tests may delay a Soldier's RTD and encourage secondary gains.

DEFER DIAGNOSIS OF BEHAVIORAL DISORDERS

8-21. During assessment, COSC personnel must always consider BH disorders that resemble COSR, but defer making the diagnosis. The COSC personnel favor this default position to preserve the Soldier's expectations of normalcy (according to BICEPS). This is also done to avoid stigma associated with BH disorders and to prevent the Soldier identifying with a patient or sick role. Deferral is also preferred because some diagnoses require extensive history collection or documentation that is unavailable during deployment situations (such as personality disorders and attention deficit hyperactive disorder). It is possible that a Soldier can have a combination of COSR, BH disorders and physical illnesses/injuries at the same time. In such cases, COSC personnel must rely on their clinical experiences, training, and consultation with peers and medical personnel to distinguish among these sometimes overlapping conditions. Physical injuries/illnesses are treated at a MTF, however, the Soldier may return for further COSC interventions and activities. Deferral of diagnosis is preferred, but diagnosis can be considered if the Soldier—

- Presents for reemerging symptoms of a previously diagnosed and/or treated BH disorder.
- Presents for refill prescription of psychotropic medication.
- Has a medical condition or BH disorder listed above in paragraph 8-19.
- Is enrolled in a reconditioning program.
- Fails to improve after having received four to five days of continuous COSC interventions and activities at hold or refer status.
- Requires individual BH treatment.
- Is referred for multiple episodes of COSR.

DIAGNOSTIC CONSIDERATIONS FOR DIFFERENTIAL DIAGNOSTIC DISORDERS

Low-Grade Environmental or Stress-Related Illnesses

8-22. Low-grade environmental or stress-related DNBI illnesses can drain the Soldier's strength and confidence. For example, chronic diarrhea and slight fever may exhaust, demoralize, and contribute to COSR among Soldiers. These conditions should be treated medically, concurrently with physical replenishment, rest, reassurance and organized activities, which restore the Soldier's confidence. If they persist in spite of rest and symptomatic treatment, a more aggressive workup and treatment may be indicated.

Dehydration

8-23. Dehydration deserves special mention because it can be very insidious. Soldiers under battlefield or heavy work conditions become extremely dehydrated without feeling thirsty. This is especially likely in CBRN equipment, or in a desert/arctic environment. Refer to FM 4-25.10 for additional information on the prevention of dehydration and FM 4-25.11 for first aid measures.

Hyperthermia

8-24. Hyperthermia (overheating) in an otherwise healthy individual often first causes mild elation and excessive energy. This may be followed by irritability, disorientation, and confusion. When core body temperature climb above 106° Fahrenheit (F) or 41° Centigrade (C), the Soldier may become belligerent, combative, and have visual hallucinations. If brain temperature rises further, the Soldier collapses and convulses in heatstroke. Refer to FM 4-25.11 for first aid measures and FM 4-25.10 for additional information on prevention and first aid measures for heat injuries.

Hypothermia

8-25. Hypothermia may cause an individual to become disoriented when core body temperature falls below 95°F (34.6°C). The person may move and speak slowly. His skin looks and feels warm, leading him to take off clothing. He becomes disoriented, then unresponsive and may appear to be dead. Hypothermia is as likely in cool or even warm, wet climates as it is in extremely cold ones. Refer to FM 4-25-10 for additional information on prevention and FM 4-25.11 for first aid measures.

Overuse Syndromes

8-26. Overuse of muscles, joints, and bones that have not been prepared for the strain of field duties can lead to persisting stiffness, pain, swelling, and orthopedic injuries. If severe, these injuries may require evacuation to a hospital for evaluation. Even if these injuries are avoided, the unfit person who overexerts has days of stiffness, aching, and weakness. Such cases are likely to develop COSR if further demands are made on them.

Rhabdomyolysis

8-27. Rhabdomyolysis is one potentially dangerous complication of severe muscle overuse (and of heatstroke or crush injuries) in which myoglobin from damaged muscle cells injures the kidneys. It can cause fatigue, seizures, muscle tenderness, and muscle aches. A warning sign is dark (tea-colored) urine, but without laboratory testing, this is not easily distinguished from the concentrated urine of dehydration. Rhabdomyolysis is a medical emergency.

Head Trauma

8-28. Concussion may stun the individual and cause amnesia, residual confusion, and/or impulsive behavior. For any case of suspected head trauma or for any case of significant memory loss (especially for a discrete period of time), check scalp, eyes, ears, nose, cranial nerve signs, and vital signs for evidence of head injury. Cases with deteriorating mental status are medical emergencies. If one pupil becomes larger than the other, it is an extreme emergency requiring immediate hospitalization. Left untreated, the condition can progress rapidly to coma and respiratory arrest within hours. If a head injury is suspected, monitor mental status and vital signs periodically, especially respiration, even though physical findings are negative. Awaken the Soldier periodically to check mental status and pupil size (allowing sufficient time to recover from any grogginess on awakening). Continuous monitoring is appropriate if there are serious concerns about the risk.

Spinal Cord Trauma

8-29. Pressure, bruising, and hematomas of the spinal cord, as well as severing of the spinal cord, can cause spinal shock, with loss of sensory and/or motor functions below the level of the injury in the affected dermatome and muscle group patterns. The loss of function may be bilateral, unilateral, or partial. These cases could be confused with paralysis or sensory loss presentations of COSR. Further manipulation of a fractured spine can worsen or make permanent the spinal cord damage. Information from the history of onset, a cautious physical and neurological examination, or complete relief of symptoms following hypnosis or strong positive suggestions could demonstrate convincingly that this is only COSR. It is best to be cautious and keep the spine immobile during care and transportation.

Postconcussion Syndromes

8-30. Postconcussion syndromes may persist weeks to months beyond the period of acute concussion. Postconcussion syndromes may include perceptual or cognitive impairment, poor impulse control, and difficulty in planning ahead. These are often accompanied by cranial nerve deficits or soft neurological signs.

Abdominal Trauma

8-31. Ruptured spleen or other intraperitoneal bleeding may cause shock. The Soldier may arrive in a fetal position and be unresponsive but have reflex guarding due to peritonitis.

Air Emboli and Focal Brain Ischemia

8-32. High blast overpressures from incoming high explosive ordnance can produce air emboli (bubbles in the blood) and focal brain ischemia (small areas in the brain which cannot get oxygen because the blood flow has been interrupted). Nuclear explosions can do this, as can high explosives when shock waves are magnified by reflection within bunkers, buildings, and trenches. Ruptured eardrums, general trauma, and evidence of pulmonary damage should be detectable. Cases may have stroke symptoms (loss of muscle strength, loss of sensation in parts of the body, and/or speech disturbances), which may be subtle and mistaken for COSR.

Laser Eye Injury

8-33. Laser range finders/target designators cause small burns on the retina if they shine directly into the eye, even at great distances and especially if viewed through optics. If the laser beam causes a small retinal blood vessel to bleed inside the eyeball, the person will see red. If blood inside the eye is confirmed on examination, the Soldier should be evacuated to a hospital with verbal reassurance that he may RTD soon. If the laser does not hit a blood vessel, the Soldier may see only flashes of light, followed quickly by some painless loss of vision. If the laser damage areas of the eye responsible for peripheral vision, the Soldier may never recognize a visual deficit. If the Soldier was looking directly at the laser source however, there will likely be a major loss of visual clarity. With simple retinal burns in the retina's periphery, most of the visual symptoms recover with hours to days of rest, reassurance, and nonspecific treatment the same as with COSR. Calm, professional treatment at each level of medical care should emphasize that the injury is not vision-threatening and the chances for some, if not total, recovery is good. Soldiers with the simple retinal burns should provide self-care to decrease the risk of assuming a patient mind-set and to promote their chances of RTD. For additional information on the management of laser eye injuries, refer to FM 8-50.

Middle Ear Injuries/Diseases

8-34. Temporary loss of hearing can be cause by a decreased acoustic sensitivity following a brief extremely intense noise (explosive) or less intense, longer duration noise. Tinnitus (ringing in the ears) can also result from acoustic nerve damage or irritation, as well as from high doses of certain drugs, such as aspirin. Distinguishing physiologic from psychogenic hearing loss may require consultation with an otolaryngologist (ears, nose, and throat specialist).

Peripheral Neuropathies

8-35. Peripheral neuropathies include compression neuropathies, which are especially likely in military settings (for example, *rucksack palsy*). Depending on severity, they may require temporary job reclassification during convalescence. As these injuries are not life-threatening, a hasty diagnosis should not precede a trial of Soldier restoration treatment.

Uncommon Endemic Neurologic Disorders

8-36. These physical diseases can first manifest with cognitive emotional and/or behavioral symptoms. A comprehensive neurological examination is required for the definitive diagnoses. Examples include—

- Guillain-Barre Syndrome manifests with muscle paralysis, usually without sensory loss, which ascends the legs and arms, then the trunk, over hours to days. It is sometimes triggered by immunizations, as might be given to troops deploying overseas. It often progresses to a life-threatening situation as the muscles of respiration become involved. This requires evacuation to EAC and continental United States (CONUS). Fortunately, recovery is usually complete, but it takes months to years.

- Multiple sclerosis is a disease that can mimic many types of COSR/BH disorders with its sometimes transitory, shifting motor, sensory, speech, and cognitive/emotional symptoms. It is made worse by stress and may be difficult to diagnose. Once confirmed, multiple sclerosis cases should be evacuated to CONUS, as should other rare, progressive diseases like Lou Gehrig's disease (amyotrophic lateral sclerosis).
- True convulsive seizure can manifest after head injury or a sublethal or chronic nerve agent exposure. These are treated with normal anticonvulsant medications. Fear of nerve agent exposure may lead some Soldiers to experience psychogenic seizures. These psychogenic seizures are also called pseudo-seizures. In addition to falling unconscious and convulsing, urinary and fecal incontinence can occur during a pseudo-seizure.

SUBSTANCE ABUSE/DEPENDENCE

ALCOHOL

8-37. Substance abuse is an example of misconduct stress behaviors and not necessarily COSR. Drug and alcohol abuse may occur in active CZs and nearby areas where use is explicitly prohibited and severely punished. Personnel performing the COSC assessment should be familiar with evaluation and treatment of substances abuse and dependency. Combat operational stress control providers should consider the following:

- Heavy habitual use of alcohol, even by otherwise capable officers and NCOs, may go unnoticed in peacetime. However, alcohol abuse it may degrade necessary mission performance demanded by combat and may result in withdrawal symptoms when access to alcohol is interrupted. Alcohol withdrawal is potential medical emergency; consultation with medical personnel is essential.
- Intoxication or withdrawal from alcohol, barbiturates, and tranquilizers may be mistaken for COSR or another BH disorder. Intoxication or withdrawal requires medical treatment. Withdrawal seizures or impending or ongoing delirium tremens need emergency medical treatment.

OVERUSE OF STIMULANTS

8-38. Stimulants may cause panic attacks, hyperactivity, mania, rage attacks, psychosis or paranoia. Cessation of amphetamines after prolonged use causes a crash characterized by extreme sleepiness, lethargy, overeating depression and suicidal thinking. This condition may require one to two weeks of hospitalization to assure safe recovery.

HALLUCINOGENIC DRUGS

8-39. Hallucinogenic drugs cause sensory distortion, panic, bizarre thoughts, and potentially dangerous behaviors. These drugs may be employed by the enemy as chemical or biological warfare agents. Phencyclidine hydrochloride (PCP) is especially problematic since it also blocks pain and tends to make those under its influence paranoid, violent, and abnormally strong. Hallucinogenic drug psychosis should not be treated with antipsychotic drugs. Physically restrain and sedate patients as necessary.

INHALATION OF FUMES

8-40. Inhalation of fumes (either by accident or as deliberate abuse) and carbon monoxide poisoning can cause disoriented, abnormal behavior. Supportive treatment and, specific antidotes/medication may be needed.

ANTICHOLINERGIC DELIRIUM

8-41. In combat, atropine delirium may occur. Soldiers are equipped with atropine injectors to use as first aid against nerve agents. Two milligrams (mg) (one atropine injector) without nerve agent challenge can cause rapid pulse, dry mouth, slightly dilated pupils, decreased sweating (hot, dry, flushed skin), and urinary retention. In some individuals, 6 mg of atropine (equal to three atropine injectors) may cause hallucination and disorientation in the absence of a nerve agent challenge. Such side effects may be more pronounced in sleep-deprived Soldiers. Overheated Soldiers are more susceptible to the atropine side effects. Atropine compounds the complications of overheating by diminishing the body's ability to lose heat through sweating. One dose (2 mg) of atropine can reduce the efficiency of heat-stressed Soldiers. Two doses (4 mg) will sharply reduce combat efficiency, and 6 mg will incapacitate troops for several hours. Some plants can also cause anticholinergic delirium when eaten.

ANTICHOLINESTERASES

8-42. A nerve agent is an anticholinesterase similar to many insecticides. Low-dose nerve agent exposure may produce miosis (pinpoint pupils) without other signs. Miosis decreases vision except in very bright light and may cause eye pain when attempting to focus. This miosis may take hours to days to improve spontaneously, depending on the degree and type of exposure. Evidence gathered from farm workers poisoned by insecticides suggests that mild personality changes, insomnia, nightmares and chronic persistent depressive symptoms may be seen even after use of an antidote. Low-dose nerve agent exposure may lower the seizure threshold of many Soldiers. True epileptic seizure cases must be distinguished from those Soldiers who may have pseudo-seizures.

BEHAVIORAL DISORDERED PATIENTS IN THE THEATER

PRIMARY BEHAVIORAL DISORDERS

8-43. Primary BH disorders (especially schizophrenic-form/schizophrenic disorder, major depression, and bipolar disorder) will continue to occur at approximately the same rate as in peacetime. Some Soldiers may hide their disorders by receiving care through civilian channels. Once in the theater they may experience a relapse or self-refer to a MTF when their medication supply is exhausted.

PERSONALITY DISORDERS

8-44. Preexisting personality disorders may make a Soldier unable to adapt to military life. However, studies have failed to show a relationship between personality disorders and the likelihood of breakdown in combat. Once Soldiers with personality disorders have developed COSR or a BH disorder, they may have greater difficulty recovering and RTD. Diagnosis should never be made in haste; diagnostic criteria must be supported with adequate historical information.

Chapter 9

Combat and Operational Stress Control Stabilization

SECTION I — EMERGENCY STABILIZATION

STABILIZATION

9-1. Emergency stabilization is the acute management of disruptive behavior resulting from COSR and/or a behavioral disorder. The disruptive behavior severely impacts unit functioning by posing a danger to self and/or others. In some cases, an underlying medical condition leads to the disruptive behavior and may present an additional threat to the Soldier's life. Emergency stabilization consists of interventions that temporarily reduce a disturbed Soldier's threat of self or others, thereby allowing further medical evaluation and/or treatment. Some behavioral disorders are associated with violent behavior, such as psychotic disorders, bipolar manic disorders, antisocial personality disorder, and borderline personality disorder. Violent behavior is also associated with disruption of brain functioning due to organic factors such as intoxication, hyperthermia, metabolic imbalance, or CBRN exposure.

9-2. The COSC triage process will be repeated throughout the emergency stabilization and will determine the disposition of the Soldier. After emergency stabilization, subsequent triage could result in an immediate RTD, transfer to a COSC Soldier restoration program for observation, or further evacuation.

METHODS USED FOR EMERGENCY STABILIZATION

9-3. Methods that may be used for emergency stabilization include—

- Providing verbal reassurance and reorientation are the best methods for controlling an agitated or disruptive Soldier. If these fail, a nonthreatening show of strength or force may suffice, or sedating medications may be offered to the Soldiers. If all other means fail to reduce the threat to self and/or others, physical restraint must be considered. Given the risk for violence, it is inadvisable to attempt subdue/restraint method one-on-one.
- Applying physical restraints is reserved for subduing and restraining agitated or disruptive Soldiers who fail to respond to safer and less restrictive forms of restraint (for example, verbal warnings or show of strength). Placing a disturbed Soldier into physical restraint increases the risk of injury to the Soldier and restraint team. Prolonged or improper application of physical restraint can cause injury to the disturbed Soldier. Given the potential for injury, it is paramount that COSC personnel receive training in proper physical restraining methods. Safe medical evacuation by ground (preferred) or air ambulance.
- Regardless of the method, the restrained Soldier must be checked frequently to guard against nerve injuries or impaired circulation, which may lead to skin ulcers or gangrene. It is important to check periodically to ensure the Soldier is not secretly escaping from restraints. The Soldier is provided verbal reassurances with positive expectations for his recovery each time he is checked.
- Chemical restraints (such as medication) can be administered to a disturbed Soldier to reduce the risk of harm to self or others. Medication may be offered to the Soldier in conjunction with verbal reassurances and reorientation. Chemical restraints may only be ordered by a medical professional who is authorized to prescribe medication when a Soldier is incompetent to make medical decisions for himself and/or when the Soldier's behavior places himself or others in danger. Once administered, medical personnel must observe for side effects and adverse reactions, and must consider administering additional medications as needed. When a Soldier is in physical restraints, medication may no longer be essential, but may serve to reduce the risk of escape, limb damage, and overheating. As a secondary benefit, once the medications reduce the

Soldier's agitation, others in the vicinity may feel safer and calmer. Before prescribing an antipsychotic medication, there are a few things to consider. First, some antipsychotic drugs may take several hours or days to take effect. Second, early administration of an antipsychotic drug may confuse the clinical picture for the next evaluator in the evacuation chain. The recommendation for most cases is to use no medication unless it is truly necessary for management.

SECTION II — COMBAT AND OPERATIONAL STRESS CONTROL FULL STABILIZATION

FULL STABILIZATION

9-4. Full stabilization is normally the mission of the Level III MTF such as a CSH specialty clinic's psychiatric service. Full stabilization goes beyond securing the safety of the Soldier and those around him. It provides a safe environment for the Soldier to receive treatment interventions, continued evaluation, and assessment for RTD potential. If RTD within the evacuation policy is not feasible, the full stabilization process helps to prepare the patient for a safe, long-distance evacuation. If a Level III MTF is not available, full stabilization may be accomplished by COSC personnel when appropriately supported or by using equipment diverted from the COSC Soldier restoration capability. Considerations for full stabilization may include—

- Conducting full stabilization for NP patients is desirable for the sake of the Soldier's future treatment and for the potential of returning some Soldiers to duty. However, full stabilization is personnel intensive with a relatively low RTD payoff. Providing only sufficient stabilization to allow evacuation from the theater may be acceptable in order to maintain the other COSC functions.
- Ensuring appropriate timely evacuation of Soldiers with NP/behavioral disorders according to the theater evacuation policy. It is preferred that full stabilization is achieved for all NP patients to facilitate appropriate and timely evacuation according to the theater evacuation policy.
- Assessing and triaging of COSR Soldiers undergoing full stabilization is an ongoing process. In subsequent triages, if a Soldier becomes stable the potential for RTD, they may be transferred to a COSC Soldier restoration or reconditioning program, or may be RTD directly.

TENETS OF FULL STABILIZATION

9-5. The COSC full stabilization includes ongoing evaluation of RTD potential. This requires assessment of mental status and performance capability overtime without excessive drug effects or limitation on activity. Contact with the Soldier's unit may be needed to get information on prior history and functioning. The further from the unit the Soldier has been evacuated, the more difficult it may be to contact the Soldier's unit. Full COSC stabilization normally takes several days. To the extent compatible with safety, the stabilization program should adhere to the principles and methods for treating COSR and behavioral disorder (such as BICEPS and 5 R's).

9-6. During full stabilization, special efforts should be made to maintain and reinforce the Soldier's identity as a Soldier. Techniques that may be helpful in maintaining the Soldier mind-set include—

- Keep Soldiers in duty uniform, not pajamas, as soon as they can be allowed.
- Maintain rank distinctions and appropriate military courtesy.
- Encourage self-care and helping behaviors.
- Engage in military work activities appropriate to the Soldiers' level of function and MOS.

9-7. Initial and ongoing assessments are essential to tailor the treatment to the Soldier's individual needs. It is essential that clinical documentation is available for these assessments. The Soldier's condition is an evolving one, and must be monitored throughout full stabilization. If assessed capable to RTD, efforts should be made to return the Soldier to duty.

9-8. Ongoing treatment and/or therapeutic modalities are essential to improving a Soldier's chances to RTD whether in theater or after evacuation. Therapeutic modalities are similar to those used on inpatient units, but must remain consistent with COSC principles. These modalities include medication, individual psychotherapy, group psychotherapy, and appropriate therapeutic occupations. Observed responses to therapeutic modalities allow informed recommendation for RTD status.

FULL STABILIZATION FACILITIES

9-9. Full stabilization is commonly conducted in the CSH. The CSH can provide more sophisticated procedures, laboratory and x-ray capabilities than are available at a medical company MTF. If a medical unit, CSC is providing emergency or full stabilization, the Soldier must be kept separate from other Soldiers in Soldier restoration or reconditioning.

9-10. If the Level III MTF (CSH) cannot provide sufficient inpatient psychiatric treatment for Soldiers requiring stabilization and preparation for evacuations, the theater/AO COSC consultant may recommend up to two temporary COA until the shortfalls resolve. He can recommend to the higher medical command that COSC personnel from one or more medical unit, CSC augment the Level III MTF psychiatric service until the caseload decrease or until replacements or additional COSC personnel are brought into theater or the MTF. Lastly, the medical C2 headquarters may direct that a COSC Soldier restoration or reconditioning assets be collocated with the Level III MTF to provide an "overflow" ward as well as augmenting staff. Definitive information is provided on the CSH NP ward staff capabilities in FM 4-02.10.

9-11. The MF2K CSH has a psychiatrist, three psychiatric nurses and a medical/surgical nurse, a social worker, nine mental health specialists and an OT sergeant at full strength. It has a modular hospital tent to provide a 20-bed hospital ward. The MF2K hospitals were designed to be at one location, but may follow the MRI doctrine of splitting off a portion to a remote site, taking with it some of the psychiatry service personnel. One portion would not have the psychiatry ward module, and would need to function similarly to the MRI CSH psychiatry service.

9-12. The MRI CSH has a psychiatrist, a psychiatric nurse, and two MH specialists. The service does not have an organic psychiatric ward where NP patients are admitted. There is normally an intermediate care ward with medical ward staff designated to receive NP patients. Some of these patients seen by the NP clinic will already be stable, and are being evaluated and prepared for RTD or evacuation. Most patients who are admitted for emergency or full stabilization will be on the designated ward, although a few with serious medical complications could require the intensive care ward. If the hospital has a high census of medical and surgical patients, most of the NP patients will need to occupy beds on the intermediate care ward. The MRI hospital may be augmented by staff from a medical company/detachment, CSC to operate a NP ward, as required.

9-13. Full stabilization facilities in theater can be categorized into two types:

- Mobile facilities use general purpose (GP) large or medium tents as used in medical units, CSC and can also be available to a CSH or a tent, expandable, modular, personnel (TEMPER) tent as used in a CSH.
- Fixed facilities use buildings that were previously hospitals or buildings converted to hospitals.

9-14. Facilities used for full stabilization include tents, Deployable Medical Systems (DEPMEDS) TEMPER tents, and fixed facilities. The adaptation of these facilities have both advantages and disadvantages that include—

- The principal advantage of the (hospital) TEMPER tents, as assembled into DEPMEDS hospital, is their climate control capability. This may be a significant safety advantage for treating seriously disturbed patients in restraints with high-dose medication, which can disrupt body's ability to regulate normal body temperature. The TEMPER and standard tents both pose greater problems for security than do fixed facilities. The staff may, therefore, have to rely more than is ideal on physical restraints and medications for sedation of some cases. Blankets or screens can be used to isolate or segregate problem patients from others. Such partitions reduce mental contagion but provide little true protection.

- Standard hospital beds are on high, lightweight metal legs with wheels. For full stabilization purposes, these should be replaced with standard low, stable cots to hold strong, agitated patients in restraints. The cots also make a more "military" setting and can be used as seats for group activities.
- When feasible, it is best to have a separate "closed" (high security) and "open" (moderate/minimal) security area. The latter could be a standard GP large tent (the same as those of the minimal care wards [MCWs] located close to the TEMPER tent of the official NP ward). The specialists (MOS 68WM6 [practical nurse] and 68W [health care specialist]) of the MCW could be given on-the-job training in supervision and military group activities for the moderate/minimum security cases if the NP staff is too small.

Chapter 10

Combat and Operational Stress Control Soldier Restoration

SECTION I — SOLDIER RESTORATION

SOLDIER RESTORATION PROGRAM

10-1. Soldier restoration is normally a 24- to 72-hour (1- to 3-day) program in which Soldiers with COSR receive treatment. Soldier restoration is accomplished using the principles of BICEPS and the 5 R's as discussed in the previous chapter. The 5 R's are tailored to the needs of the Soldier. Soldier restoration is conducted as close to the Soldier's unit as possible. Soldier restoration can be conducted by medical units/elements throughout the theater with the assistance of organic and/or augmented COSC personnel. The medical company/detachment, CSC is staffed and equipped to establish Soldier restoration programs. Soldier restoration may also be accomplished at a Level III MTF (CSH). The tenets remain the same but implementation may differ in the duration of Soldier restoration and the specialized skill and knowledge of available providers. There are three subcategories of Soldier restoration (referred to as lines of Soldier restoration) based on location and available resources that are discussed in Section II of this chapter.

SOLDIER RESTORATION PROCESS

10-2. The process of Soldier restoration involves several steps that include screening, assessment, and interventions, reintegration/coordination, and movement of Soldiers. A discussion of each of the steps in the process is provided below.

SCREENING

10-3. Adequate medical screening and treatment as necessary must be done before the Soldier begins the Soldier restoration program. Minor medical conditions can be treated during routine sick call. Soldiers entering a Soldier restoration program should be only those *hold* cases that require continuous medical and/or BH evaluation and observation for 24 hours or more. The criteria for *hold* cases are discussed under COSC triage in Chapter 8. Tending to and restoring physiological status (such as sleep and hydration) is a priority. Some Soldiers who need Soldier restoration will also have a mental disorder. Treatment for the mental disorder may continue or be initiated during the Soldier restoration process.

ASSESSMENT AND INTERVENTION

10-4. Initial assessment and subsequent COSC interventions depends upon the severity of the COSR and to what extent the reactions interfere with the Soldier's ability to function. More thorough interviews are conducted only after the Soldier's physiological status has been restored.

REINTERGRATION

10-5. Coordination is required to assist Soldiers with reintegration back into their units. Soldiers are reassessed regularly. When a Soldier experiencing COSR begins to improve, the COSC interventions shift toward the reintegration of the Soldier back into his unit. The COSC personnel must work with the COSC consultant of their area or other resources to assist with this reintegration.

MOVEMENT

10-6. Movement of Soldiers from one line of Soldier restoration to another that has greater capabilities or security may be required. When movement of these Soldiers is required, vehicles used to transport them should be nonambulance vehicles, if possible, and accompanied by escorts from their unit. If nonmedical unit members are not available to perform escort duties, medical augmentation may be required.

PRINCIPLES AND PROCEDURES OF SOLDIER RESTORATION

10-7. Initial Soldier restoration begins as close to the Soldier's unit as possible, normally near an MTF where the Soldier can get a reprieve from extreme stress but at the same time can be close to his unit. Normally, Soldier restoration facilities are established in the BSA near or adjacent to the FSMC/BSMC Level II MTF. Soldier restoration is not feasible at locations that are consistently under artillery, air, or direct-fire attack, but ideally are still within the sound of the artillery or other reminders of battle. If there is potential for attack, there must be cover and defensive positions. The location should not be one from which a move is likely within 24 hours. If there is a significant possibility of a move, only those cases that can participate actively in the move with minimal supervision should be managed at this location. The specific site of the Soldier restoration facility should be adjacent to the medical company's area and be out of the immediate (close) sight of the medical triage and treatment areas

REASSURANCE

10-8. Immediate reassurance is given to the Soldier with COSR beginning with the COSC triage as discussed in Chapter 8. Tell the Soldiers that they are temporarily joining the unit, not as patients, but as Soldiers who need a few days to recover from COSR. Emphasize that COSR is a normal response to extremely abnormal conditions. Rapid recovery is also normal and RTD is expected. Reassure the Soldiers about safety and what to do in the event of an attack or march order. Lastly, orient the Soldiers to the Soldier restoration program.

STRUCTURED MILITARY ENVIRONMENT

10-9. An emphasis should be placed on maintaining a highly structured military unit environment and schedule of activities in order to keep the Soldier from adopting a patient role. Assign the Soldier to a squad under supervision of a specific squad leader. The squad leader may be a CSC unit/MH section member or a member of the BSB subordinate unit. The squad leader may also be one of the Soldiers there for Soldier restoration, if their condition allows. The newly arrived Soldier is assigned to a tent that has been designated/erected for the Soldier restoration program. Soldiers with more dramatic COSR or BH symptoms should be temporarily quartered separately from other Soldiers receiving Soldier restoration. In remaining consistent with the principle of treating Soldiers with COSR as Soldiers instead of patients, it is recommended that Soldier restoration facilities do not display the Red Cross as displayed on the MTF. This could affect their protection under the Geneva Conventions. See FM 4-02 for information on the Geneva Conventions.

SUPPORT THE SOLDIER'S MILITARY IDENTITY

10-10. Maintain appropriate rank distinctions, titles, and military courtesies within the confines of the tactical situation. Expect the Soldier to maintain military bearing and personal appearance. They should be in duty uniform. Conduct basic Soldiering skills. Do not take personal possessions away from the Soldier. This includes weapons, unless there is significant concern for the Soldier's safety. If the Soldier arrives with a weapon, SOPs of the AO will guide whether or not the weapon is secure for the Soldier. Also, encourage the unit to maintain contact with its Soldier.

REPLENISHMENT OF PHYSIOLOGICAL STATUS

10-11. Get the Soldier under shelter and cool down if overheated, warm up if cold, and dry off if wet. Providing hot beverages and/or soup will also assist with restoring body temperature. Replenish hydration with palatable beverages and meals. Unless the Soldier is totally exhausted, institute some personal hygiene. If the Soldier needs uniforms or equipment, coordinate with the BSB S4 or supporting logistic element. Restorative sleep should be as normal as possible. Soldiers are typically able to fall asleep when reassured of safety, and do not routinely require medication to induce sleep. They should be informed that they may awaken with vivid and frightening dreams, and be instructed on quick relaxation techniques to go back to sleep. If those measures are insufficient, one-time medication with a sedative/hypnotic may be considered, coordinated, and prescribed by a physician or PA from the supporting medical company. The duration of sleep should be sufficient to make substantial progress in repaying the sleep debt. It should also begin the process of restoring a reasonable sleep/wake cycle that is consistent with the Soldier's duties in his unit.

RESTORE CONFIDENCE

10-12. Restoring a Soldier's confidence may include—

- Providing regularly scheduled formations to keep Soldiers informed of daily activities and the tactical situation, including information about their unit, when available.
- Providing therapeutic occupations that are based on the Soldier's current functional ability. The OT (or COSC team members under the guidance of the OT) selects therapeutic occupations that support the Soldier's military identity and enhance the Soldier's sense of competence. Therapeutic occupations may include activities of daily living (such as physical fitness or uniform maintenance); educational activities (selected common and collective Soldier task training or life skills training); work/productive activities (militarily relevant tasks such as vehicle maintenance or site maintenance); leisure/recreational activities (enjoyable, relaxing activities, games, and sports); social participatory activities (cooperative/competitive sports, games, ceremonies, or celebrations). Therapeutic occupations provide the casualty with a challenge but afford successful performance that shows the Soldier that he is still capable and competent. This realization plays the dominant role in restoring the Soldier's sense of confidence, functional capacity, and ability to RTD.
- Training and teaching Soldier on methods for managing excessive stress.
- Facilitating factional review of precipitating event that has caused the Soldier's COSR through talking (ventilating) and coaching. Combat and operational stress control personnel (or health care providers trained by COSC personnel) facilitate these discussions. They help Soldier restore perspective with questions and coaching. The factual review (debriefing) is often best done individually unless several Soldiers experienced the same event.

SECTION II — LINES OF SOLDIER RESTORATION

FIRST-LINE SOLDIER RESTORATION

10-13. First-line Soldier restoration is usually provided at the BSMC/FSMC in the BSA. It is provided by personnel organic to the BSMC/FSMC usually assisted and supervised by COSC personnel organic to the medical company and/or from CSC unit personnel. This also applies to a cavalry regiment. Because of the high mobility of the BSMC/FSMC, Soldier restoration will often be a 24- to 72-hour process. First-line Soldier restoration also may occur in relative secure locations in division, corps, and theater AOs. First-line Soldier restoration in the division is provided near and supported by a Level II MTF such as a MSMC with its organic MH and/or minimal augmentation from a CSC unit. First-line Soldier restoration in corps and theater is provided by ASMCs with responsibility for their AO. It is provided by COSC personnel organic to the medical company and/or from CSC unit personnel. The MSMC and ASMCs will be relatively unlikely to move on short notice so Soldier restoration up to 72 hours should be feasible. Support provided for Soldier restoration is a responsibility of the MH sections of supporting medical companies. Tasking for support of Soldier restoration programs is accomplished through the command surgeon.

SECOND-LINE SOLDIER RESTORATION (SOLDIER RESTORATION CENTER)

10-14. Each of the first-line Soldier restoration locations listed above should be backed up by a second-line Soldier restoration capability at a location that is less likely to have to move on short notice and has more COSC capabilities. Soldier restoration centers may be located near the MSMC, ASMC, or with a CSC medical company or detachment. They can be located in the BSA near the BSMC/FSMC when it is at a stable base camp. The Soldier restoration center may receive *hold* cases that are transferred from the first-line medical companies, as well as Soldiers from nearby units. Second-line Soldier restoration has more equipment and a greater range of COSC expertise. This permits a 72-hour holding capacity for a stable well-organized Soldier restoration center and may provide full stabilization.

THIRD-LINE SOLDIER RESTORATION

10-15. In some scenarios, units with Soldiers in need of Soldier restoration may be significantly closer to a Level III MTF such as a CSH than to a medical company or CSC unit. On order, a Soldier restoration program may be conducted by the CSH specialty clinic NP staff, which may also be augmented with personnel from a medical company/medical detachment, CSC. In such cases, a Soldier restoration program may be conducted by the CSH specialty clinic NP staff. It may be augmented with personnel from the medical company or detachment, CSC. Soldier restoration near a CSH must be kept clearly separate from the patient ward. It may be done at a MCW or in separate tents.

RETURN TO DUTY OF RECOVERED COMBAT AND OPERATIONAL STRESS REACTION SOLDIERS

10-16. Most COSR symptoms do not necessarily improve completely while the prospect of combat continues. The Soldiers should be given the positive expectation that they will RTD. Every possible effort should be made to return Soldiers to their original unit. The RTD of Soldiers following Soldier restoration depends on how near the Soldier's unit is, the availably of a means of transportation, and the tactical situation. Ideally, their units are contacted to send someone to get the Soldier or he may be returned to his unit by way of the personnel replacement company. Mental health personnel coordinate directly with the unit to which the Soldier is returning and/or with COSC, other medical personnel, or UMT supporting the unit's AO. These contacts can consult with the leaders of the Soldier's unit and facilitate the Soldier's acceptance back into his unit.

DOCUMENTATION

10-17. Soldiers receiving Soldier restoration interventions must be tracked from the initial contact until they are returned to their unit. A record must be maintained of interventions and activities provided during Soldier restoration, as well as the Soldier's response. A notation of the dates, any pertinent medical data, and providing unit is entered in the Soldier's medical record. An administrative summary of Soldier restoration services may be developed, stored, and disposed of as directed by the Soldier restoration center SOP. If the Soldier received BH treatment, documentation should be maintained as delineated in AR 40-66. A statistical record is maintained and sent to higher medical headquarters according to SOP.

Chapter 11

Behavioral Health Treatment

SECTION I — BEHAVIORAL HEALTH CARE

TREATMENT FOR BEHAVIORAL DISORDERS

11-1. Behavioral health treatment exists when there is an explicit therapist-patient or therapist-client relationship. Behavioral health treatment is provided for Soldiers with behavioral disorders to sustain them on duty or to stabilize them for referral/transfer. This is usually brief, time-limited treatment as dictated by the operational situation. Behavioral health treatment includes counseling, psychotherapy, behavior therapy, occupational therapy, and medication therapy. Treatment assumes an ongoing process of evaluation, and may include assessment modalities such as psychometric testing, neuropsychological testing, laboratory and radiological examination, and COSC providers' discipline-specific evaluations.

COMBAT AND OPERATIONAL STRESS CONTROL BEHAVIORAL HEALTH PROVIDER

11-2. There are five professional disciplines and two enlisted specialties that serve as BH/COSC providers, see Chapter 3. Army BH/COSC providers are officers who hold active individual clinical privileges granted by an MTF according to AR 40-68. Providers also include enlisted medical specialist who do not hold individual clinical privileges, but who work under direct supervision of a professional COSC provider. Behavioral health treatment is provided within the scope of the providers' clinical privilege, training, and experience.

SECTION II — BEHAVIORAL HEALTH TREATMENT PROTOCOLS AND MEDICATIONS

BEHAVIORAL HEALTH TREATMENT FOR SOLDIERS

11-3. Behavioral health treatment is provided to Soldiers with diagnosed behavioral disorders (see Chapter 8), and who require more intentions for their diagnoses. It is both inappropriate and detrimental to treat Soldiers with COSR as if they are a BDP. A therapeutic relationship may promote dependency and foster the "patient" role. Likewise, medication therapy and the highly structured treatment modalities imply the "patient" role. Medication for transient symptom relief (insomnia or extreme anxiety) may not be detrimental if there is no expectation that medication will continue to be prescribed.

BEHAVIORAL HEALTH PROTOCOLS AND PROCEDURES FOR TREATMENT

11-4. Behavioral health treatment is provided within a context of preventing symptoms of behavioral disorder from progressing in severity and improving the level of functioning. The principles of BICEPS apply. Treatment should not impede activities for other prevention levels (for example, universal, selective, and indicated). Treatment should not interfere with the Soldier's duties or the unit's mission. The COSC provider should emphasize Soldier's identity as a Soldier despite having a diagnosis. Transportation of the Soldier to the provider (or vice versa) and ability to assure reliable meeting times can be limiting factors, depending on the operational environment. Medication refills must be coordinated for availability.

Standards of Treatment

11-5. Treatment standards are the same in the deployed environment as in garrison. When operational requirements dictate that clinical standards of treatment/care are waived or relaxed, it must be approved by the AO COSC consultant. (AR 40-68 provides additional information.) Treatment should be tailored to the anticipated availability of the Soldier and COSC provider. Short-term interventions are more practical than long-term commitments. If longer-term treatment is necessary, design the intervention in time-limited modules. Under no circumstances should treatment diminish the Soldier's ability to provide self-care and to defend himself. Exceptions include emergency stabilization and preparation for evacuation. In addition, the VHA/DOD Clinical Practice Guidelines website (http://www.oqp.med.va.gov/cpg/cpg.htm) offer clinicians evidence-based assessment and treatment algorithms for acute stress disorder, PTSD, and many other behavioral disorders.

Army Regulations Governing Evaluations

11-6. Fitness for duty evaluations are conducted as necessary within the priorities of the supported commanders according to AR 40-501; psychiatrists should not initiate a medical evaluation board without first ensuring the Soldier has received adequate treatment. This treatment may not be available in theater. Command-directed evaluations are conducted as necessary within the priorities of the supported commanders according to DODD 6490.1 and MEDCOM Regulation 40-38. Clinical documentation should be safeguarded according to AR 40-66 and local command policy. Treatment should be conducted in a location that is as private as possible. Information can be released to a third party if the Soldier consents. Combat and operational stress control providers need to notify command when the Soldier's safety is in question (suicidal, homicidal) or if the Soldier is removed from his unit for medical observation. Additional release of information to command is on a need-to-know basis. For any questions on release of information on COSR and NP patients, consultation with supporting JAG office is advised. Also, maintaining the Soldiers health record with clinical data is required and accomplished according to AR 40-66 and supplemental theater policy as appropriate. Treatment should occur throughout the evacuation process and follow-up is expected at home station.

Chapter 12
Reconditioning

SECTION I — RECONDITIONING PROGRAM

LOCATION

12-1. Reconditioning programs are intensive efforts to restore those Soldiers triaged as a *refer* case, but who still have good potential for RTD. Referral to reconditioning can be from Level II MTF, COSC Soldier restoration program, or from Level III MTF. Reconditioning includes the rehabilitation of Soldiers with mental disorders, such as substance abuse/dependency. Reconditioning programs are conducted by COSC personnel and Soldiers usually participate for up to 7 days. Soldier participation may be extended by a case-by-case exception to theater evacuation policy. Reconditioning is conducted only in a theater where there are adequate COSC elements and supplies. When COSC resources are needed for other functional areas, any ongoing reconditioning program reduces its scope of services or closes. Reconditioning is provided at first-, second-, or third-line centers determined by location and available resources.

PROVIDERS

12-2. Conducting reconditioning programs is a mission of the medical company, CSC and the medical detachment, CSC (MRI). Reconditioning can also be done on a small-scale by a CSC medical detachment (MF2K) under some circumstances.

SECTION II — TENETS AND PROCEDURES OF RECONDITIONING

RECONDITIONING

12-3. Reconditioning may be considered an extension of Soldier restoration. Reconditioning is similar to Soldier restoration, but with potentially longer stay, treatment strategies focus on preventing atrophy of skills and assisting Soldiers in regaining skills needed for duty. Reconditioning makes more use of BH treatment modalities.

STRUCTURED MILITARY ENVIRONMENT TO SUSTAIN SOLDIER'S IDENTITY AS A SOLDIER

12-4. Like Soldier restoration, reconditioning emphasizes a highly structured military unit environment and schedule of activities in order to keep the Soldier from adopting a "patient" role. Maintaining a military environment is even more critical when colocated with hospitals or other service support elements.

REPLENISHMENT OF PHYSIOLOGIC STATUS AND CONFIDENCE

12-5. Reconditioning initially emphasizes physical replenishment and hygiene, but later shifts the emphasis to more closely match the conditions that the Soldier should expect when RTD. In order to restore confidence in the Soldier, unit formations are held on a regularly scheduled basis, and include the Soldiers receiving reconditioning.

OCCUPATIONAL THERAPY

12-6. Therapeutic occupations may include—

- Activities of daily living (for example, hygiene, physical fitness, and uniform maintenance).

- Educational activities (for example, common and collective Soldier task training, and life skills training).
- Militarily relevant work details according to the current functional level and MOS/duties of the Soldier.
- Leisure/recreational activities (for example, enjoyable, relaxing activities, including cooperative physical and mental activities, and basic relaxation techniques).
- Social participatory activities (for example, games and ceremonies).
- Group training in relaxation techniques.
- Ventilation and coaching where COSC (or medics trained by COSC personnel) encourage discussion about stressors and their impact on mental well-being. This technique helps the Soldier to restore personal perspective with questions and coaching. Individual counseling and therapy may improve the Soldier's functions as well.

12-7. When reclassification recommendations are considered, an adjacent CS or CSS unit may be able to provide a job that will match the Soldier's abilities. This provides an opportunity to demonstrate the Soldier's abilities and build confidence. The reconditioning personnel may recommend to S1/G1 that the Soldier be reclassified to another MOS.

FIRST-LINE RECONDITIONING PROGRAM

12-8. First-line reconditioning programs in the corps are staffed by task-organized CSC elements from the CSC company or by an MRI CSC detachment. Reconditioning can be conducted by an MF2K CSC detachment, but would preempt its Soldier restoration capability. If the inpatient NP workload is light, a small reconditioning center may be staffed by personnel from the NP ward and consultation service of an MF2K CSH. The first-line reconditioning center usually collocates with a CSH but must maintain its separate, nonhospital identity. It should not be situated among the hospital wards, or near the morgue, triage area, or helicopter pad. Within a theater/AO, the preferred option is to have one or two reconditioning centers placed such as to allow easy access from Soldier restoration centers. Under major, prolonged combat conditions, it may be more appropriate to have one reconditioning center behind each division. The reconditioning center is dependent on the hospital for support. The reconditioning center works with the supporting hospital by sending work parties of Soldiers in the program to assist in food preparation and delivery and cleanup chores. These Soldiers may also be used for assisting with work details throughout the hospital, but must be under direct supervision of either hospital or BH personnel. The reconditioning center uses the supporting hospital's medical records section to maintain the permanent case records. Cases in the reconditioning center are counted as patients in the reconditioning center on the daily hospital census. The cases are not counted as occupied beds when reporting the hospital bed occupancy. Upon disposition from the reconditioning center, whether for RTD, retraining for other duty, or evacuation, the reconditioning center prepares the chart for further evacuation or writes the discharge summary and closes the hospital's chart.

12-9. Some reconditioning cases will be able to return to far forward CS or CSS duty. However, many of the Soldiers who need reconditioning will be unable to return to their original unit, due to combat operations or the nature of their symptoms. Soldiers who undergo a 4- to 7-day Soldier restoration program in the corps should not be crossed off the division's personnel rolls (as specified in AR 40-216). Continuing contact with the unit increases RTD rates. For Soldiers who recover but are no longer on the division's rolls, every feasible effort should be made to return them to their original units. The chief, division MH and the CSC unit teams that are attached to reinforce the division should maintain frequent contact with the reconditioning centers that support the division. Coordination efforts, through the division support command (DISCOM) sustainment brigade and the personnel replacement system by division BH personnel, should attempt to facilitate return of recovered Soldiers to their original units. If recovered Soldiers cannot be returned to their previous small unit, consider forming them into cohesive pairs or small groups which can be reassigned to a new unit together. The standard corps evacuation policy is seven days, but some Soldiers with good potential for RTD may need a few more days at the reconditioning center. The corps evacuation policy is at the discretion of the theater commander. The principle purpose of a short evacuation policy is to keep hospital beds available for mass casualties and to minimize the expense and labor required to treat serious surgical and medical cases.

SECTION III — EVACUATION POLICY

EXTENDING THE THEATER EVACUATION POLICY

12-10. If need assessment indicates that a longer period of time (more than seven days) will achieve a Soldiers' RTD, the theater COSC consultant can request permission to extend the theater evacuation policy (up to 14 days). He submits the request for authority via the commander, MEDBDE, through the corps surgeon to the corps commander. The following facts apply: The reconditioning program, because of its austerity, is not a significant logistical burden to the corps. Soldiers in reconditioning perform useful work details and perimeter defense. Lastly, the increase in RTD is important in operations such as stability and reconstruction operations, where reducing personnel attrition becomes an important factor.

EVACUATION

12-11. Reconditioning cases that do not recover sufficiently to return to some duty within the designated evacuation period are evacuated from corps to EAC. They are best transported in cargo trucks and buses, bus ambulances, or an ambulance train. In the latter two cases, they should be assigned helper tasks. Use air evacuation only if there is no other alternative.

SECOND-LINE RECONDITIONING

12-12. Second-line reconditioning is conducted in the intermediate staging base outside the CZ in the theater. This center could be at a CSH, a fixed MTF, or a CSC unit element. The second line reconditioning center continues to emphasize physical fitness, Soldier skills, work details, and individual/group counseling/psychotherapy. Cases will be retrained for CSS duties at corps and theater levels. As soon as the recovering Soldiers are ready, the retraining site can shift to on-the-job training at a nearby CSS unit.

THIRD-LINE RECONDITIONING

12-13. Third-line reconditioning is for Soldiers with COSR and/or mental disorders who did not improve sufficiently at lower lines of reconditioning but still have RTD potential. These Soldiers are transferred to a reconditioning program at a regional MEDCEN or home station MEDDAC. These reconditioning centers, like the others, must maintain a military atmosphere and provide opportunities to engage in occupational therapy. Some Soldiers may also require retraining for other duty/MOS reclassification.

Appendix A
Combat and Operational Stress Control and Religious Support

ROLE OF UNIT MINISTRY TEAM

A-1. This appendix addresses the general role of the UMT in the commander's program for COSC. The UMT, imbedded within units down to battalion level, provide immediate support to leaders in fulfilling their COSC responsibilities. The UMTs also assist in training leaders to recognize combat and operational stress identification and intervention responsibilities. In cooperation with unit medical personnel, UMTs serve as a primary referral agency to BH resources.

RELIGIOUS SUPPORT FOR COMBAT AND OPERATIONAL STRESS CONTROL

A-2. Soldiers' inner resources are generally rooted in their religious and spiritual values. In combat, Soldiers often show more interest in their religious beliefs. When religious and spiritual values are challenged by the chaos of combat, Soldiers may lose connection with the inner resources that have sustained them. The UMT is the primary resource available to Soldiers experiencing such dilemmas and is a valuable resource in assisting them as they seek to refocus their spiritual values.

Unit Ministry Team Support for Combat and Operational Stress Control

A-3. The UMT provides preventive, immediate, and restorative spiritual, emotional support and care to Soldiers experiencing COSR.

Preventive Religious Support for Combat and Operational Stress Reaction

A-4. The UMT assists in preventing COSR and misconduct stress disorders through spiritual fitness training. Ministry of presence with Soldiers, assigned Department of Army civilians, and contractors is critical. The UMT provides a stabilizing influence on personnel and assists them in strengthening and regaining personally held spiritual values. Preventative activities include—

- Worship opportunities.
- Private and group prayer opportunities.
- Religious literature and materials.
- Scripture readings.
- Sacraments and ordinances.
- Assistance to personnel and families prior to deployment, emphasizing family strengths.

Immediate Religious Support for Combat and Operational Stress Reaction

A-5. The UMT assists commanders in the identification of personnel experiencing negative reactions to combat and operation stress, COSR, and misconduct stress behaviors. The UMT works closely with the unit's leaders and medical personnel to care for COSR cases through religious support and comfort. Immediate religious support activities may include—

- Conversation focused upon fears, hopes, and other feelings.
- Prayer for fallen comrades and memorial ceremonies and services.
- Rites, sacraments, and ordinances, as appropriate.
- Sacred scripture.

Restorative Religious Support for Combat and Operational Stress Control

A-6. Following an operation, a unit may require reconstitution. Surviving Soldiers may need to rebuild emotional, psychological, and spiritual strength. Depending upon the spiritual, emotional, and physical condition of the unit's Soldiers, the organic UMT may need augmentation from higher echelons or other units. Restorative religious support activities may include—

- Worship, sacraments, rites, and ordinances.
- Memorial ceremonies and services.
- Religious literature and materials.
- Grief facilitation and counseling.
- Reinforcement of the Soldiers' faith and hope.
- Opportunities for Soldiers to talk about combat experiences and to integrate those experiences into their lives.

Appendix B

Medical Detachment, Combat and Operational Stress Control

CONCEPT OVERVIEW

BACKGROUND

B-1. The current medical units, CSC in the Army inventory were initially developed over ten years ago. The MF2K redesign resulted in the development of the MRI. Recent events in the Global War on Terrorism in both Afghanistan and Iraq have indicated a need to update these units from MRI medical detachment, CSC to a multifunctional COSC detachment. The 2003 US Army Surgeon General's Mental Health Advisory Team (MHAT) report and observations indicate a need to update the unit in terms of modularity, multifunctionality, and its working relationship with support units and command surgeons from brigade to theater levels.

PROPOSED MEDICAL DETACHMENT, COMBAT AND OPERATIONAL STRESS CONTROL

B-2. The proposed medical detachment, COSC (TOE 08660G000) consists of a detachment headquarters, a main support section, and a forward support section. The main support section consists of its headquarters and an 18-Soldier BH team made up of social workers, clinical psychologists, psychiatrists, occupational therapists, psychiatric nurses, MH specialists, and OT specialists. The forward support section consists of an 18-Soldier BH team only. Each BH team is capable of breaking into six 3-person subteams, for battalion/company prevention and fitness support activities. This provides for a total of 12 subteams for each detachment, giving supported commanders more teams and more flexibility in the utilization of those teams, yet maintaining all of the capabilities of COSC.

Operational Environment

B-3. Soldiers will continue to experience the physical and mental impacts of high stress in both combat and stability and reconstruction operational environments. Advances in technology which will increase effectiveness, and the impact of the individual Soldier, will also increase dispersion and add to isolation and stress levels of Soldiers. The enemy of the future will look for methods that will have the greatest psychological impact on our future Soldiers; future enemies will not be our technological equals and will increasingly rely on terror and nontraditional methods to unnerve, injure, and demoralize both Soldiers and civilians. There is an identified need to update COSC units in terms of flexibility, modularity, multifunctionality, and its relationship to supported units and the surgeons responsible to commanders for the BH mission. The current configuration of the COSC units resulted in an uneven use of available COSC resources and that there was a greater need for preventive and limited fitness services forward. Commanders have communicated there is a greater need for active outreach and consultation.

Assumptions

B-4. Power projection will likely remain the fundamental strategic and operational imperative of our forces for the foreseeable future. The new medical detachment, COSC will be well suited to the future force, designed to move with and support the division and corps of the future force, and division and corps equivalents. The medical detachment, COSC will be assigned to a MEDBDE, MMB, or MEDCOM and will be operationally attached to supported units.

B-5. Each BCT has a MH section with one BH officer and one MH specialist and that all division and brigade MH sections are filled. The medical detachments, COSC will be able to augment BCTs MH sections. The medical detachment, COSC will require additional logistical, finance, maintenance, personnel, legal, FHP, and administrative services support.

Implication for Combat and Operational Stress Control

B-6. Future forces must be emotionally and mentally fit before deploying and resilient in battle in order to endure the fast pace and intense lethality of the battlefield. Commanders will need to be able to apply COSC capabilities at every level quickly and effectively. Reorganization of the medical detachment, CSC to a combined, multifunctional medical detachment, COSC will provide greater flexibility and greatly increase outreach and preventive services while maintaining capability for the entire spectrum of COSC.

Operational Concept, Mission

B-7. The medical detachment, COSC may be attached to a theater or division medical C2 headquarters in order to provide COSC casualty prevention, treatment, and management on an area basis.

B-8. The AUTL (see FM 7-15) linkage includes—

- ART 6.5. Provide Force Health Protection in a Global Environment.
- ART 6.5.1. Provide Combat Casualty Care.
- ART 6.5.1.5. Provide Mental Health/Neuropsychiatric treatment.
- ART 6.5.4. Provide Casualty Prevention.
- ART 6.5.4.5. Provide Combat Operational Stress Control Prevention.

Method of Employment

B-9. The medical detachment, COSC deploys with a division-sized force into a theater or with a task force of up to 25,000 Soldiers in stability and reconstruction operations. The medical detachment, COSC is attached for OPCON to the division/corps/theater medical C2 headquarters. The medical detachment, COSC is assigned to the MMB of units that have reorganized under the future force design. It is normally tasked by the command surgeon to support units within the command. The detachment commander as authorized by his medical C2 headquarters will coordinate operations with the command surgeon or the division/corps/theater psychiatrist. In the absence of a psychiatrist, the detachment commander will advise and consult with the command surgeon and the commander as required. Subteams may be task-organized and further attached for operational control to maneuver BCTs, and will operate in and forward from BSA, working closely with medical assets and combat religious support team (CRST). The detachment commander will coordinate efforts with the division and the CRST to provide in-theater prevention programs as required (suicide prevention, reunion, COSC training, TEM, and PTE stress management).

Detachment Headquarters

B-10. Plans, supervises and monitors the activities of the medical detachment, COSC in support of the COSC mission. See Table B-1 for a listing of personnel assigned to the detachment headquarters. The commander will establish a command post (CP) collocated with the supported unit surgeon (either division or sustainment brigade) and coordinate all activities of the detachment through the surgeon section. The detachment commander provides C2 and works closely with supported unit surgeon to provide logistics and other required services for the detachment. The field medical assistant and/or the detachment sergeant also establishes and coordinates logistics and required services support through the supported command staff for all detachment personnel and operations.

Table B-1. Detachment headquarters personnel

Detachment headquarters personnel
Detachment Commander (LTC/05, AOC 05A00)
Medical Operations Officer (CPT/03, AOC 70B67, MS)
Chaplain (CPT/03, AOC 56A00, CH)
Detachment Noncommissioned Officer in Charge (NCOIC) (SFC/E7, MOS 68X40)
Supply Sergeant (SGT/E5, MOS 92Y20)
Human Resources Specialist (SPC/E4, MOS 42A10)
Wheeled Vehicle Mechanic (SPC/E4, MOS 63B10)
Patient Administration Specialist (SPC/E4, MOS 68G10)
Religious Support Specialist (SPC/E4, MOS 56M10)
Cook (PFC/E3, MOS 92G10)
Note. The position of detachment commander may be filled by a (LTC, AOCs 73A00, 73B00, 60W00, 65A00 or 66C00). For description of duties, see Section I of the TOE and Department of Army (DA) Pamphlet 611-21.

Main Support Section

B-11. Included in the main support section is the detachment headquarters personnel identified in Table B-1 above and an 18-personnel BH team. See Table B-2 for list of personnel assigned to the BH team. This BH team can break into 6 three-person subteams, each with one officer, one NCO, and one specialist to conduct prevention and limited fitness operations in the battalion/company level in a maneuver BSA. Teams are collocated with a BCT medical unit. Modularized subteams may be combined and task-organized to provide or support any other COSC function, which includes larger scale Soldier restoration reconstitution, or any other COSC or BH mission as determined by its headquarters. Tasks for the main support section BH team includes:

- Consultation and coordination with commanders at team level (battalion and company).
- Coordination with CRST.
- Prevention and fitness support activities.
- Reconstitute any other COSC function as required.

B-12. When attached for support to any medical company Level II MTF, the grouped subteams are dependent on the supporting medical company for logistical support and quarters. The subteam can provide 3-day restoration for up to 50 Soldiers. With additional logistical support and quarters from supported command, the team can accommodate additional Soldiers under surge conditions.

Table B-2. Main section behavioral health team

Main section behavioral health team
Team Leader/Psychiatrist (MAJ/04, AOC 60W00)
Occupational Therapist (MAJ/04, AOC 65A00)
Social Work Officer (MAJ/04, AOC 73A67)
Behavioral Science Officer (CPT/03, AOC 67D00)
Clinical Psychologist (CPT/03, 73B67)
Psychiatric Nurse (CPT/03, 66C7T)
Occupational Therapy NCO (Four)—(Two SSGs/E6 and Two SGTs/E5, MOS 68W30/68W20)
Mental Health NCO (Four)—(One SSG/E6 and Three SGTs/E5, MOS 68X30/68X20)
Mental Health Specialist (Six)—(SPC/E4, MOS 68X10)
Note. For description of duties, see Section I of the TOE and DA Pamphlet 611-21.

Forward Support Section

B-13. This section consists of an 18-Soldier BH team. It is comprised of a section leader and five other BH officers, six MH/OT NCOs, and six MH specialists. This BH team can break into six three-person subteams, each with one officer, one NCO, and one specialist to conduct prevention and limited fitness operations in the BCT medical company in the BSA. Modularized subteams may be combined and task-organized to provide or support any other COSC function, which includes larger scale Soldier reconstitution, or any other COSC or BH mission as determined by its headquarters. The forward support section is dependent on the supported unit for C2 and logistical support. See Table B-3 for list of personnel assigned to the forward support section BH team.

Table B-3. Forward support section behavioral health team

Team Leader/Psychiatrist (MAJ/04, AOC 60W00)
Psychiatric Nurse (MAJ/04, 66C7T)
Clinical Psychologist (MAJ/04, 73B67)
Occupational Therapist (CPT/03, AOC 65A00)
Social Work Officer (CPT/03, AOC 73A67)
Behavioral Science Officer (CPT/03, AOC 67D00)
Occupational Therapy NCO (Two)—(One SSG/E6 and One SGT/E5, MOS 68W30/68W20)
Mental Health NCO (Four)—(One SSG/E6 and Three SGTs/E5S, MOS 68X30/68X20)
Mental Health Specialist (Six)—(SPC/E4, MOS 68X10)
Note. For description of duties, see Section I of the TOE and DA Pamphlet 611-21.

Organizational Concept

B-14. Required capabilities require the unit to—

- Deploy to power projection platform or point of debarkation and deploy with supported unit or deploy to theater and join supported unit.
- Provide all functions of COSC from company to theater level in close coordination with and under the direction of the supported unit surgeon.
- Provide prevention, assessment, treatment, and referral BH activities.
- Provide consultation with commanders from company to theater level.
- Conduct and coordinate administrative and logistical support to sustain operations.

Basis of Allocation Total Army Analysis

B-15. The basis of allocation is 0.333 per BCT; 1 per division; 2 per theater. In support of a theater, a COSC detachment provides support on an area basis and provides additional support to the division/corps on order. See Figure B-1 for breakdown of separate standard requirement codes (SRCs) of the medical detachment, COSC sections.

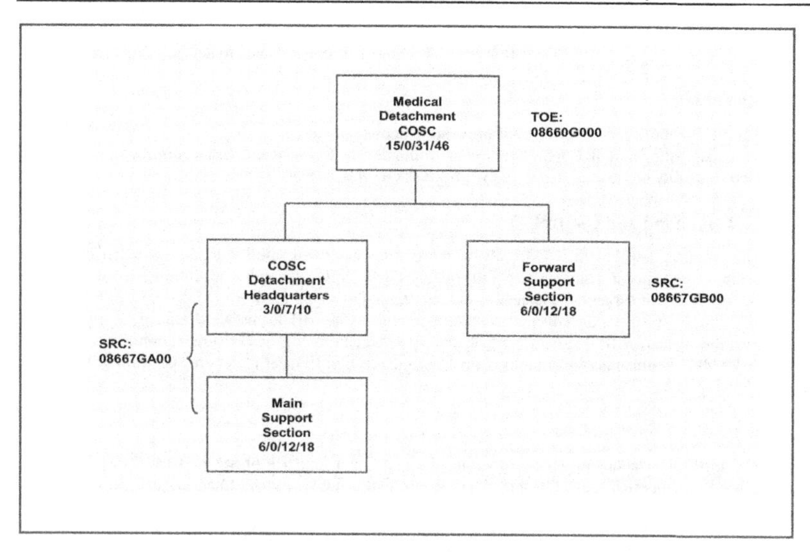

Figure B-1. Medical detachment, combat and operational stress control

Concept of Transition

B-16. Transition begins with the four active component detachments, one at a time. Personnel, transportation, and equipment changes will be minimal; transition should take a short amount of time and money to accomplish. Changes in the detachments are congruent with Force Modernization designs and personnel proponency issues. The proposed COSC detachment is designed with the Future Force in mind.

Patient Medical Records

B-17. The Surgeon General's MHAT report notes that there is a need to standardize BH reporting and documentation. The BH charts were inconsistently maintained and documentation did not always accompany patients through the evacuation chain. An additional finding was that there was no standardized method of collecting BH workload or clinical data and that no database that tracked evacuees provided reliable clinical information. The patient administration specialist (MOS 68G10) will supervise or perform functions as outlined in DA Pamphlet 611-21 for the Soldier caseload; maintain Soldier/patient accountability and records; interface with patient administration division of supported/collocated medical units.

Role of Religious Support Specialist

B-18. Findings of the MHAT include the need for chaplains to be aware of their role in COSC and that an aggressive chaplain outreach program should be executed. This requires that supervisory chaplains be involved in CRST integration with BH and primary care providers, and educate primary players in role of CRST in COSC. These activities require that the COSC chaplain be mobile and have an active presence in the maneuver brigades. This position is required to fulfill the full-time force protection requirements for the chaplain while in theater. The religious support specialist assists the COSC chaplain by screening individuals seeking counseling, coordinates for convoy security when the CRST goes out on site visits; assists chaplain with fund request; develops and coordinates needed religious support projects within the unit; coordinates with detachment for all vehicle services, communication system repairs, detachment training requirements and section accountability. This Soldier is cross-trained to develop skills for COSC.

Multifunctionality

B-19. The MH sections have a priority of prevention activities, but also have the capability to conduct or support any combat operational stress function as required. The COSC teams will be outfitted to a standard to gain the capability of prevention and limited fitness activities.

Modularity, Command and Control

B-20. The medical detachment, COSC attached early to supported units will have a better-integrated relationship. Detachment commanders and NCOICs will collocate with and coordinate all activities through supported unit surgeons. Detachments are designed to support a division in combat or a joint task force of up to 25,000 personnel in noncombat operations. The BH teams are designed to support and operate within maneuver BCTs in support of maneuver battalions for area coverage, collocating in BSA with the BSMC. Teams may be combined for task-organization to meet the entire spectrum of COSC as required.

Limitations

B-21. The medical detachment, COSC does not have its own logistics or services capability and must depend on the supported unit for logistical, legal, FHP, food service, finance, and personnel and administrative services. The COSC teams are small (one or two vehicles) and are dependent upon supported units for convoy and security operations. The logistical support necessary (space, tents, cots, and so forth) to provide Soldier restoration/reconstitution functions must be furnished by the supported element (division, corps, or theater).

Assumptions

B-22. Wartime activities will continue to cause physical and mental impacts of extremely high stress in both direct combat and the support operational environment. Advances in technology will impact the individual Soldier by increasing dispersion, and adding to their isolation and stress levels. The enemy of the future will look for methods that will have the greatest psychological impact on our future Soldier; future enemies will not be our technological equals and will increasingly rely on terror and nontraditional methods to unnerve, injure, and demoralize both Soldiers and civilians. Commanders will need to be able to quickly and effectively apply COSC capabilities at every level of the unit.

Appendix C
Medical Company and Medical Detachment, Combat Stress Control (Medical Force 2000)

SECTION I — MEDICAL COMPANY, COMBAT STRESS CONTROL (TOE 08467L000) (MEDICAL FORCE 2000)

MISSION

C-1. The mission of the AOE medical company, CSC is to provide comprehensive COSC support through directed interventions activities and COSC training for supported corps units. It provides DS to maneuver brigades lacking organic BH officers; augments units with BH assets; and provides area support. The medical company, CSC reconstitutes other COSC assets. The medical company, CSC provides COSC interventions and activities to indigenous populations as directed in stability and reconstruction operations, to include domestic support operations, humanitarian assistance, disaster relief, peace support operations, and detention facility operations. The medical company, CSC provides COSC interventions and activities to units in support of their readiness preparation and throughout their deployment cycle.

ASSIGNMENT

C-2. The AOE medical company, CSC (TOE 08467L000) is assigned to a corps MEDBDE (TOE 08422A100) or corps MEDCOM (TOE 08411A000).

EMPLOYMENT

C-3. Company headquarters usually locates with the corps medical headquarters. The employment of its teams includes their dispersal throughout the corps AO. Teams may be attached to an ASMB; a division or brigade medical company; a CSH; or other corps medical unit headquarters.

CAPABILITIES

C-4. At TOE Level 1, the medical company, CSC includes—

- Dividing the preventive section into six mobile COSC preventive teams.
- Dividing the restoration section into four mobile restoration teams with each team being equipped to hold 40 Soldiers at the same time. With additional logistical support, each team can accommodate additional Soldiers under surge conditions.
- Assisting higher headquarters BH staff with planning and coordination of COSC support, identifying the stress threat and mental and physical stressors, and the implementation of COSC functional areas.

DEPENDENCY

C-5. This unit is dependent on—

- Appropriate elements of the corps for legal; finance; field feeding; personnel and administrative services support; laundry and clothing exchange; mortuary affairs support; and security of EPW, detainee, and US prisoner patients.
- The medical headquarters to which it is assigned/attached for FHP; medical administration; logistics (including MEDLOG); medical regulating of patients; evacuation; coordination for RTD; and unit-level equipment and CE maintenance.

Mobility and Security

C-6. This unit is 100-percent mobile. It requires 100 percent of its organic personnel and equipment be transported in a single lift, using its organic vehicles. Upon relocation, Soldiers being held will require additional transportation. This unit is responsible for perimeter defense of its immediate operational area. However, it is dependent on appropriate elements of the corps for additional security, to include security of convoy operations. Personnel of the company (except the chaplain) are provided weapons for their personal defense and for the defense of their patients and/or held Soldiers.

Organization

C-7. The AOE medical company, CSC (Figure C-1) is organized into a headquarters section, a preventive section with six (modular) CSCP, and a restoration section with four (modular) combat stress control restoration (CSCR) teams. The medical company, CSC is only present in the MF2K and is an RC organization that is programmed to be in the Army inventory until the year 2010. Under the MRI a new medical detachment, CSC, TOE 08463A000, will replace this unit.

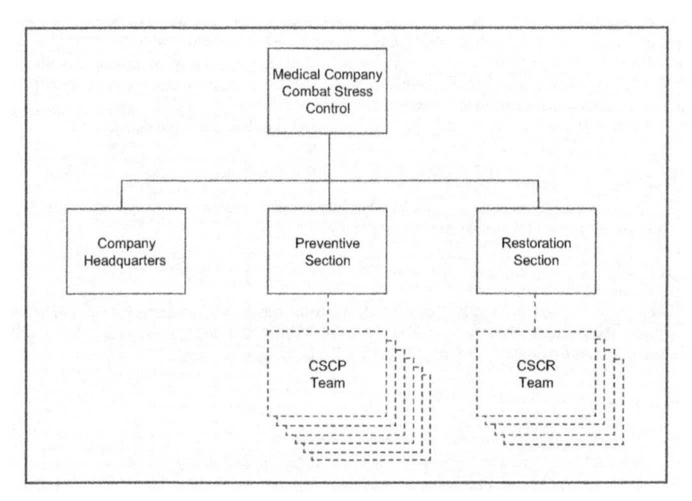

Figure C-1. Medical company, combat stress control

Company Headquarters Section

C-8. The company headquarters section provides C2 and unit-level administrative and maintenance support to its subordinate sections when they are collocated with the company. Personnel from the headquarters section are deployed with teams or task-organized COSC elements, as required. Personnel assigned to this section are identified in Table C-1.

Table C-1. Company headquarters section personnel 3/0/14

Company Commander (LTC/05, AOC 60W00, MC)
Chaplain (CPT/03, AOC 56A00, CH)
Field Medical Assistant (CPT/03, AOC 70B67, MS)
First Sergeant (1SG/E8, MOS 68W5M)
Mental Health NCO (SFC/E7, MOS 68X40)
Motor Sergeant (SSG/E6, MOS 63B30)
Supply Sergeant (SGT/E5, MOS 92Y20)
Equipment Record/Parts Sergeant (SGT/E5, MOS 92A20)
Wheeled Vehicle Mechanic (SGT/E5, MOS 63B20)
Human Resource Specialist (SPC/E4, MOS 42A10)
Nuclear, Biological, and Chemical Specialist (SPC/E4, MOS 74D10)
Wheeled Vehicle Mechanic (SPC/E4, MOS 63B10)
Unit Supply Specialist/Armorer (SPC/E4, MOS 92Y10)
Power-Generation Equipment Repairer (SPC/E4, MOS 52D10)
Cook (Three)— (Two SPC/E4 and One PFC/E3, MOS 92G10)
Note. The cook is deployed to assist the unit providing food service support to the company. He may participate in other CSC unit missions including outreach surveillance and stress control training.

Preventive Section

C-9. The 24-person preventive section staff is identified in Table C-2. The preventive section is task-organized to conduct COSC interventions and activities. Preventive section personnel may be task-organized with personnel of the COSC restoration section into teams for specific missions. The preventive section can divide into six CSCP teams. The section (and team) leader position may be held by any of the officers assigned to the section. The preventive section can augment or reconstitute the medical detachment, CSC teams.

Table C-2. Preventive section personnel 12/0/12

Psychiatrist (Three)—(MAJ/04, AOC 60W00)
Social Worker (Two)—(MAJ/04, AOC 73A67)
Psychiatrist (Three)—(CPT/03, AOC 60W00)
Social Worker (Four)—(CPT/03, AOC 73A67)
Team Chief (Six)—(SGT/E5, MOS 68X20)
Mental Health Specialist (Six)—(SPC/E4, MOS 68X10)

Restoration Section

C-10. The 44-person restoration section staff is identified in Table C-3. The restoration section may be task-organized to perform its COSC mission. Combat stress control restoration section personnel may also be task-organized with personnel of the CSCP section into teams for specific missions. This section can be divided into four CSCR teams. Each CSCR team can deploy a four-person mobile team using their HMMWV.

Table C-3. Restoration section personnel 12/0/32

Restoration section personnel
Occupational Therapist (MAJ/04, AOC 65A00)
Psychiatric/Mental Health Nurse (Two)—(MAJ/04, AOC 66C00/7T)
Clinical Psychologist (MAJ/04, AOC 73B67)
Occupational Therapist (Three)—(CPT/03, AOC 65A00)
Psychiatric/Mental Health Nurse (Two)—(CPT/03, AOC 66C00/7T)
Clinical Psychologist (Three)—(CPT/03, AOC 73B67)
Mental Health NCO (Two)—(SFC/E7, MOS 68X40)
Occupational Therapy NCO (SSG/E6, MOS 68W30/N3)
Team Chief (Four)—(SSG/E6, MOS 68X30)
Patient Administration NCO (SGT/E5, MOS 68G20)
Occupational Therapy Sergeant (Two)—(SGT/E5, MOS 68W20/N3)
Mental Health NCO (Four)—(SGT/E5, MOS 68X20)
Patient Administration Specialist (Three)—(Two SPC/E4 and One PFC/E3, MOS 68G10)
Occupational Therapy Specialist (Five)—(SPC/E4, MOS 68W10/N3)
Mental Health Specialist (Ten)—(Six SPC/E4 and Four SPC/E3, MOS 68X10)

SECTION II — MEDICAL DETACHMENT, COMBAT STRESS CONTROL (TOE 08463L000) (MEDICAL FORCE 2000)

MISSION

C-11. The mission of the AOE MF2K medical detachment, CSC is to provide COSC interventions and activities to supported units in its AO. It augments division and brigade MH sections; provides direct support to combat brigades without organic BH officers; and provides area support in its AO. The medical detachment, CSC reconstitutes other brigade and division COSC assets. The medical detachment, CSC provides COSC interventions and activities to indigenous populations as directed in stability and reconstruction operations, to include domestic support operations, humanitarian assistance, disaster relief, peace support operations, and detention facility operations. The medical detachment, CSC provides COSC interventions and activities between deployments to units in support of their readiness preparation and throughout their deployment cycle. Under MRI this detachment will be replaced with medical detachment, CSC, TOE 08463A000.

Note. The MF2K and MRI medical detachments, CSC conduct COSC support operations very similarly. However, in the MRI medical detachment, CSC, psychiatry assets are assigned to the restoration section and psychology assets are assigned to the preventive section whereas in the MF2K detachment, the manning strategy for psychology and psychiatry is reversed: psychologists are in the restoration section and psychiatrists are on the preventive teams.

ASSIGNMENT

C-12. The MF2K medical detachment, CSC (TOE 08463L000), is assigned to a corps MEDBDE (TOE 08422A100), or a corps MEDCOM (TOE 08411A000), or other medical task force C2 elements. Its teams may be attached to a medical company, CSC (TOE 08467L000); an ASMB (TOE 08456A000); a brigade or divisional medical company; or a CSH.

EMPLOYMENT

C-13. Detachment headquarters is usually located in the DSA. Its teams disperse and are employed throughout its AO.

CAPABILITIES

C-14. The MF2K medical detachment, CSC provides—

- Mobile COSC interventions from the preventive section which can divide into three preventive teams.
- Holding for 40 Soldiers by the restoration team for COSC interventions such as restoration. With additional logistical support, the team can accommodate additional Soldiers under surge conditions.

STAFF RESPONSIBILITIES

C-15. The medical detachment assists the C2 headquarters (to which it is assigned or attached) regarding planning and coordination of COSC support, stress threat, mental and physical stressors, stress behaviors, principles of COSC, and implementation of COSC functional areas.

DEPENDENCY

C-16. This unit is dependent on—

- Appropriate elements of the supporting unit for FHP; religious support; legal; finance; field feeding; personnel and administrative services support; laundry and clothing exchange; mortuary affairs support; and security of EPW, detainee, and US prisoner patients.
- Supporting unit for medical administration; logistics (including MEDLOG); medical regulating; evacuation; coordination for RTD; and unit-level equipment and CE maintenance.

MOBILITY AND SECURITY

C-17. This unit is 100-percent mobile. It requires 100 percent of its organic personnel and equipment be transported in a single lift, using its organic vehicles. Upon relocation, Soldiers being held will require additional transportation. This unit is responsible for perimeter defense of its immediate operational area. However, it is dependent on appropriate elements of the supporting unit for additional security, to include security of convoy operations. Personnel of the detachment are provided weapons for their personal defense and for the defense of their patients and/or held Soldiers.

ORGANIZATION

C-18. This 25-person unit (see Table C-4) is composed of a headquarters, a combat stress preventive section with three preventive teams, and a CSCR. The modular CSC teams found in the MF2K CSC medical detachment are similar to those found in the CSC medical company.

Table C-4. Detachment teams

Headquarters	*Preventive Section*	*Restoration Section*
	CSCP#1	CSCR
	CSCP#2	
	CSCP#3	

C-19. The detachment headquarters provides C2 for the detachment. It is responsible for planning, coordinating, and implementing COSC interventions and activities for supported units. It consists of three personnel: a detachment Commander (05, 60W00, MC), a detachment sergeant (E7, 68X40), and a wheeled-vehicle mechanic (E4, 63B10). The detachment commander also serves as a treating physician with the preventive section. The detachment NCOIC (a senior MH NCO) also serves as the restoration team sergeant. Detachment officers and NCOs from the prevention team and the restoration team may be assigned additional duties, which enhance the overall effectiveness of the headquarters section. See Table C-5 for personnel assigned to detachment headquarters.

Table C-5 Detachment headquarters personnel

Commander (LTC/05, AOC 60W00)
Detachment Sergeant (SFC/E7, MOS 68X40)
Wheeled Vehicle Mechanic (SPC/E4, MOS 63B10)
Note. The commander also serves as the psychiatrist in the preventive section.

Preventive Section

C-20. The 12-person preventive section staff is identified in Table C-6. The preventive section may be task-organized to conduct its COSC mission. Preventive section personnel may also be task-organized with personnel of the restoration section into teams for specific missions. The preventive section can divide into three preventive teams. The section (and team) leader position may be held by any of the officers assigned to the section. This section's COSC interventions and activities are—

- Unit needs assessment; consultation and education; critical event and transition management; COSC triage; stabilization (emergency); and BH treatment.
- Assisting with restoration and reconditioning at the CSC detachment program. Overseeing a 1- to 3-day COSC restoration program in a brigade, division, or ASMC holding section or in another area suitable for Soldiers experiencing COSR and/or other stress-related disorders.

Table C-6. Preventive section 7/0/6

Psychiatrist (LTC/05, AOC 60W00)
Clinical Psychologist (Three)—(MAJ/04, AOC 73B67)
Social Worker (Three)—(CPT/03, AOC 73A67)
Team Chief (Three)—(SGT/E5, MOS 68X20)
Mental Health Specialist (Three)—(SPC/E4, MOS 68X10)
Note. The psychiatrist is also counted in the headquarters section.

Restoration Team

C-21. The 9-person restoration team staff is identified in Table C-7. The restoration team is task-organized to provide COSC interventions and activities. Restoration team personnel may also be task-organized with personnel of the CSC preventive section into teams for specific missions. Each restoration team can deploy a four-person mobile team with a HMMWV. The CSCF usually collocates with a supported divisional medical company to provide mobile COSC support within a DSA and conduct restoration programs, as required. The CSCF provides staff and equipment for operating a restoration or reconditioning center. This section's COSC interventions and activities are—

- Unit needs assessment; consultation and education; critical event and transition management; COSC triage; stabilization; and BH treatment.
- Conducting Soldier restoration and reconditioning programs.
- Assisting the CSH psychiatric section when psychiatric ward capability is required (refer to FM 8-10-14).

Table C-7. Restoration team 3/0/6

Psychiatrist (MAJ/04, AOC 60W00)
Psychiatric/Mental Health Nurse (MAJ/04, AOC 66COO/7T)
Occupational Therapist (CPT/03, AOC 65A00)
Occupational Therapy NCO (SSG/E6, 68W30/N3)
Team Chief (SSG/E6, MOS 68X30)
Occupational Therapy Sergeant (SGT/E5, MOS 68W20/N3)
Mental Health NCO (SGT/E5, MOS 68X20)
Mental Health Specialist (Two)—(One SPC/E4 and One PFC/E3, MOS 68X10)

Glossary

SECTION I — ACRONYMS AND ABBREVIATIONS

AAR	after-action review
ABCA	American, British, Canadian, and Australian
ACR	armored cavalry regiment
AMEDD	Army Medical Department
AN	Army Nurse Corps
AO	area of operations
AOC	area of concentration
AOE	Army of Excellence
AR	Army regulation
ART	Army tactical task
ASMB	area support medical battalion
ASMC	area support medical company
AUTL	Army Universal Task List
BAS	battalion aid station
BCT	brigade combat team
BDP	behavioral disordered patient
BF	battle fatigue
BH	behavioral health
BICEPS	brevity, immediacy, contact, expectancy, proximity, and simplicity (See Section II for definition.)
BSA	brigade support area
BSB	brigade support battalion
BSMC	brigade support medical company
C	Centigrade
C2	command and control
C4I	command, control, communications, computers, and intelligence
CBRN	chemical, biological, radiological, and nuclear
CE	communications-electronics
CH	chaplain
CHL	combat health logistics
CHS	combat health support
COA	course(s) of action
CONUS	continental United States
COSC	combat and operational stress control (See Section II for definition.)
COSR	combat and operational stress reaction (See Section II for definition.)

CP	command post
CPT	captain
CRST	combat religious support team
CS	combat support
CSC	combat stress control
CSCF	combat stress control fitness
CSCP	combat stress control preventive
CSCR	combat stress control restoration
CSH	combat support hospital
CSR	combat stress reaction
CZ	combat zone
DA	Department of the Army
DD	Department of Defense
DEPMEDS	Deployable Medical Systems
DISCOM	division support command
DMHS	division mental health section
DNBI	disease and nonbattle injury(ies)
DOD	Department of Defense
DODD	Department of Defense Directive
DODI	Department of Defense Instruction
DS	direct support
DSA	division support area
EAB	echelons above brigade
EAC	echelons above corps
EAD	echelons above division
EMT	emergency medical treatment
EPW	enemy prisoner(s) of war
ETOD	end-of-tour debriefing
F	Fahrenheit
FAP	Family Advocacy Program
FHP	force health protection
1LT	first lieutenant
1SG	first sergeant
FM	field manual
FMC	United States Field Medical Card (Department of Defense Form 1380)
FRAGO	fragmentary order
FSB	forward support battalion
FSMC	forward support medical company
G-1	Assistant Chief of Staff (Personnel)
G-2	Assistant Chief of Staff (Intelligence)
G-3	Assistant Chief of Staff (Operations and Plans)

G-4	Assistant Chief of Staff (Logistics)
G-5	Assistant Chief of Staff (Plans and Policy)
G-9	Assistant Chief of Staff (Civil-Military Affairs)
GP	general purpose
HBCT	heavy brigade combat team
HHD	headquarters and headquarters detachment
HIP	help-in-place
HMMWV	high-mobility multipurpose wheeled vehicle
HSL	health service logistics
HSS	health service support
IED	improvised explosive device
JAG	Judge Advocate General
LTC	lieutenant colonel
MAJ	major
MC	Medical Corps
MCW	minimal care ward
MEDBDE	medical brigade
MEDCEN	medical center
MEDCOM	medical command (US Army)
MEDDAC	medical department activity
MEDLOG	medical logistics
METT-TC	mission, enemy, terrain and weather, troops and support available, time available, and civil considerations
MF2K	Medical Force 2000
mg	milligram(s)
MH	mental health
MHAT	Mental Health Advisory Team (US Army)
MMB	multifunctional medical battalion
MOS	military occupational specialty
MP	military police
MRI	Medical Reengineering Initiative
MRO	medical regulating officer
MS	Medical Service Corps
MSB	main support battalion
MSMC	main support medical company
MTF	medical treatment facility
MTOE	modification table of organization and equipment
NATO	North Atlantic Treaty Organization
NCO	noncommissioned officer
NCOIC	noncommissioned officer in charge
NP	neuropsychiatric

OEF	Operation Enduring Freedom
OIF	Operation Iraqi Freedom
OPCON	operational control
OPLAN	operation plan
OPORD	operation order
OPTEMPO	operational tempo
OSAT	operational stress assessment team
OSR	operational stress reaction
OT	occupational therapy
PA	physician assistant
PCP	phencyclidine hydrochloride
PD	psychological debriefing
PERSTEMPO	personnel tempo
PIES	proximity, immediacy, expectancy, and simplicity
POC	point of contact
PTE	potentially traumatizing event
PTSD	posttraumatic stress disorder
pub	publication
PVNTMED	preventive medicine
R&R	rest and relaxation
RC	Reserve Component
RTD	return to duty
S1	Adjutant (US Army)
S2	Intelligence Officer (US Army)
S3	Operations and Training Officer (US Army)
S4	Supply Officer (US Army)
SBCT	Stryker Brigade Combat Team
SGT	sergeant
SOI	signal operation instructions
SOP	standing operating procedure
SP	Army Medical Specialty Corps
SPC	specialist
SRC	standard requirement code
SSG	staff sergeant
TC	training circular
TDA	table of distribution and allowances
TEM	traumatic events management
TEMPER	tent, expandable, modular, personnel
TG	technical guide
TOE	table(s) of organization and equipment
TRADOC	United States Army Training and Doctrine Command

TSOP	tactical standing operating procedures
UCMJ	Uniform Code of Military Justice
UMT	unit ministry team
UNA	unit needs assessment
US	United States
USAMEDDC&S	United States Army Medical Department Center and School
USAMRMC	United States Army Medical Research and Materiel Command
USAR	United States Army Reserve
USMC	United States Marine Corps
VHA	Veterans Health Adminstration

SECTION II — TERMS

brevity, immediacy, contact, expectancy, proximity, and simplicity

An acronym used for the management of combat and operational stress reactions—*brevity* (usually less than 72 hours); *immediacy* (as soon as symptoms are evident); *contact* (chain of command remains directly involved in the Soldiers recovery and return to duty), *expectancy* (combat stress control unit personnel expectation that casualties will recover); *proximity* (of treatment at or as near the front as possible); *simplicity* (the use of simple measures such as rest, food, hygiene, and reassurance). Also known as BICEPS.

combat and operational stress control

A coordinated program for the prevention of and actions taken by military leadership to prevent, identify, and manage adverse combat and operational stress reactions in units. Also known as COSC.

combat and operational stress reaction

The expected, predictable, emotional, intellectual, physical, and/or behavioral reactions of Service members who have been exposed to stressful events in combat or military operations other than war. Also known as COSR.

5 R's

Actions used for combat and operational stress reaction control that include—**R**eassure of normality; **R**est (respite from combat or break from the work); **R**eplenish bodily needs (such as thermal comfort, water, food, hygiene, and sleep); **R**estore confidence with purposeful activities and contact with his unit; **R**eturn to duty and reunite Soldier with his unit.

reconditioning program

An intensive 4- to 7-day program (may be extended by exception to theater evacuation policy) of replenishment, physical activity, therapy, and military retraining for combat and operational stress control casualties and neuropsychiatric cases (including alcohol and drug abuse) who require successful completion for return to duty or is evacuated for further neuropsychiatric evaluation.

Soldier restoration

A 24- to 72-hour (1- to 3-day) program in which Soldiers with combat and operational stress reactions receive treatment.

stabilization

The initial short-term management and evaluation of severely behaviorally disturbed Soldiers caused by an underlying combat and operational stress reaction, behavioral health disorder, or alcohol and/or drug abuse reaction.

References

SOURCES USED

These are the sources quoted or paraphrased in this publication.

DEPARTMENT OF DEFENSE

These documents are available online at: http://www.dtic.mil/whs/directives/

DODD 6490.1, *Mental Health Evaluations of Members of the Armed Forces*, 1 October 1997

DODD 6490.2, *Comprehensive Health Surveillance*, 21 October 2004

DODD 6490.5, *Combat Stress Control (CSC) Program*, 23 February 1999

DODI 6490.4, *Requirements for Mental Health Evaluations of Members of the Armed Forces,* 28 August 1997

MULTISERVICE PUBLICATION

FM 6-22.5/MCRP 6-11C/NTTP 1-15M, *Combat Stress*, 23 June 2000

ARMY PUBLICATIONS

These publications are available online at: https://akocomm.us.army.mil/usapa

AR 40-216, *Neuropsychiatry and Mental Health,* 10 August 1984

FM 7-15, *The Army Universal Task List*, 31 August 2003

FM 22-51, *Leaders' Manual for Combat Stress Control,* 29 September 1994

DOCUMENTS NEEDED

These documents must be available to the intended users of this publication.

DEPARTMENT OF DEFENSE FORMS

These forms are available online at: https://akocomm.us.army.mil/usapa

DD Form 1380, *US Field Medical Card*

DD Form 2795, *Pre-Deployment Health Assessment*

DD Form 2796, *Post-Deployment Health Assessment*

MULTISERVICE PUBLICATION

AR 190-8/OPNAVINST 3461.6/AFJI 31-304/MCO 3461.1, *Enemy Prisoners of War, Retained Personnel, Civilian Internees, and Other Detainees,* 1 October 1997

ARMY PUBLICATIONS

These publications are available online at: https://akocomm.us.army.mil/usapa

AR 25-1, *Army Knowledge Management and Information Technology Management*, 15 July 2005

AR 40-3, *Medical, Dental, and Veterinary Care,* 12 November 2002

AR 40-61, *Medical Logistics Policies*, 28 January 2005

AR 40-66, *Medical Record Administration and Health Care Documentation,* 20 July 2004

AR 40-68, *Clinical Quality Management,* 26 February 2004

AR 40-400, *Patient Administration,* 12 March 2001

AR 40-501, *Standards of Medical Fitness,* 1 February 2005

AR 340-21, *The Army Privacy Program,* 5 July 1985

DA Pamphlet 611-21, *Military Occupational Classification and Structure*, 31 March 1999

FM 4-02, *Force Health Protection in a Global Environment,* 13 February 2003

FM 4-02.6, *The Medical Company—Tactics, Techniques, and Procedures,* 1 August 2002 (Change 1, 9 April 2004)

FM 4-02.7, *Health Service Support in a Nuclear, Biological, and Chemical Environment—Tactics, Techniques, and Procedures,* 1 October 2002

FM 4-02.10, *Theater Hospitalization,* 3 January 2005

FM 4-02.16, *Army Medical Information Management—Tactics, Techniques, and Procedures,* 22 August 2003

FM 4-02.17, *Preventive Medicine Services,* 28 August 2000

FM 4-25.12, *Unit Field Sanitation Team,* 25 January 2002

FM 8-10-6, *Medical Evacuation in a Theater of Operations—Tactics, Techniques, and Procedures,* 14 April 2000

READINGS RECOMMENDED

These sources contain relevant supplemental information.

JOINT OR MULTISERVICE PUBLICATIONS

Most joint publications are available online at: http://www.dtic.mil/doctrine/

Joint Pub (JP) 1-02, *Department of Defense Dictionary of Military and Associated Terms,* 12 April 2001 (As amended through 9 June 2004)

JP 3-0, *Doctrine for Joint Operations*, 9 September 2001

JP 4-02, *Doctrine for Health Service Support in Joint Operations*, 30 July 2001

FM 1-02/MCRP 5-12A, *Operational Terms and Graphic*, 21 September 2004

FM 4-25.11/NTRP 4-02.1.1/AFMAN 44-163(I)/MCRP 3-02G), *First Aid,* 23 December 2002 (Change 1, 15 July 2004)

FM 8-9/NAVMED P-5059/AFJMAN 44-151V1V2V3, *NATO Handbook on the Medical Aspects of NBC Defensive Operations AMedP-6(B), Part I—Nuclear, Part II—Biological, Part IIII—Chemical,* 1 February 1996

FM 8-284/NTRP 4-02.23(NAVMED P-5042)/AFMAN (I) 44-156/MCRP 4-11.1C, *Treatment of Biological Warfare Agent Casualties,* 17 July 2000 (Change 1, 8 July 2002)

FM 8-285(4-02.285)/NAVMED P-5041/AFJMAN 44-149/FMFM 11-11, *Treatment of Chemical Agent Casualties and Conventional Military Chemical Injuries,* 22 December 1995

FM 21-10/MCRP 4-11.1D, *Field Hygiene and Sanitation,* 21 June 2000

ARMY PUBLICATIONS

Most of these publications are available online at: https://akocomm.us.army.mil/usapa

AR 71-32, *Force Development and Documentation—Consolidated Policies*, 3 March 1997

AR 635-200, *Active Duty Enlisted Administrative Separations*, 6 June 2005

FM 3-0, *Operations*, 14 June 2001

FM 3-06.11, *Combined Arms Operations in Urban Terrain*, 28 February 2002

FM 3-07, *Stability Operations and Support Operations*, 20 February 2003

FM 3-19.40, *Military Police Internment/Resettlement Operations*, 1 August 2001

FM 3-21.21, *The Stryker Brigade Combat Team Infantry Battalion*, 8 April 2003 (Change 1, 31 July 2003)

FM 3-21.31, *The Stryker Brigade Combat Team*, 13 March 2003

FM 3-90.3, *The Mounted Brigade Combat Team*, 1 November 2001

FM 5-0, *Army Planning and Orders Production*, 20 January 2005

FM 4-0, *Combat Service Support*, 29 August 2003

FM 4-02.4, *Medical Platoon Leaders' Handbook—Tactics, Techniques, and Procedures*, 24 August 2001 (Change 1, 18 December 2003)

FM 4-02.19, *Dental Service Support in the Theater of Operations*, 1 March 2001

FM 4-02.21, *Division and Brigade Surgeons' Handbook (Digitized)—Tactics, Techniques, and Procedures*, 15 November 2000

FM 4-02.24, *Area Support Medical Battalion— Tactics, Techniques, and Procedures*, 28 August 2000

FM 4.25.10, *Field Hygiene and Sanitation*, 20 June 2002

FM 8-10-14, *Employment of the Combat Support Hospital Tactics, Techniques, and Procedures*, 29 December 1994 (will be revised as FM 4-02.14)

FM 8-42, *Combat Health Support in Stability Operations and Support Operations*, 27 October 1997 (will be revised as FM 4-02.42)

FM 8-50, *Prevention and Medical Management of Laser Injuries*, 8 August 1990 (will be revised as FM 4-02.50)

FM 8-55, *Planning for Health Service Support,* 9 September 1994 (will be revised as FM 4-02.55)

FM 10-27-4, *Organizational Supply and Services for Unit Leaders,* 14 April 2000 (will be revised as FM 4-20.05)

FM 27-1, *Legal Guide for Commanders.* 13 January 1992 (will be revised as FM 1-04.1)

FM 27-10, *The Law of Land Warfare,* 18 July 1956 (Reprinted with basic including Change 1, 15 July 1976) (will be revised as FM 1-04.10)

FM 63-20, *Forward Support Battalion*, 26 February 1990 (will be revised as FM 4-93.20)

FM 63-21, *Main Support Battalion*, 7 August 1990 (will be revised as FM 4-93.21)

FM 100-9, *Reconstitution*, 13 January 1992 (will be revised as FM 4-100.9)

FM 100-17, *Mobilization, Deployment, Redeployment, and Demobilization,* 28 October 1992 (will be revised as FM 3-35)

FM 100-17-5, *Redeployment*, 29 September 1999 (will be revised as FM 3-35)

MEDCOM Regulation 40-38, *Command-Directed Mental Health Evaluation*, 1 September 2001

TC 25-20, *A Leader's Guide to After-Action Reviews,* 30 September 1993

TRADOC Pamphlet 600-22, *Leaders Guide for Suicide Prevention Planning,* 16 February 2005 (Available at http://www.tradoc.army.mil/publications.htm)

United States Army Center for Health Promotion and Preventive Medicine

Most of these publications are available online at: http://chppm-www.apgea.army.mil/dhpw/

USACHPPM TG 240, *Combat Stress Behaviors,* June 2004

USACHPPM TG 241, *Combat Operational Stress Reaction* (*COSR*) (*"Battle Fatigue"*), June 2004

USACHPPM TG 242, *Combat Operational Stress Reactions* (*COSR*) *Prevention: Leader Actions,* June 2004

USACHPPM TG 243, *Combat Stress Card,* May 1999

Tables of Organization and Equipment

TOE 08411A000, *Medical Command, Corps* (*MRI*)

TOE 8422A100, *Medical Brigade, Corps* (*MRI*)

TOE 08456A000, *Area Support Medical Battalion*

TOE 08457A000, *Area Support Medical Company*

TOE 08463A000, *Medical Detachment, Combat Stress Control* (*MRI*)

TOE 08463L000, *Medical Detachment, Combat Stress Control* (*MF2K*)

TOE 08467L000, *Medical Company, Combat Stress Control* (*MF2K*)

TOE 08660G000, *Medical Detachment, Combat Operational Stress Control* (*Force Development Update*)

DEPARTMENT OF VETERANS AFFAIRS

Veterans Health Administration, *Clinical Practice Guidelines (Mental Health),* available online at: http://www.oqp.med.va.gov/cpg

Index

References are to paragraph numbers except where otherwise specified.

Made in the USA
Monee, IL
20 May 2022

96784795R00184